AS UNIT 1

STUDENT GUIDE

T0187376

CCEA

Physics

Unit 1

Forces, energy and electricity

Ferguson Cosgrove

HODDER
EDUCATION
AN HACHETTE UK COMPANY

Hodder Education, an Hachette UK company, Blenheim Court, George Street,
Banbury, Oxfordshire OX16 5BH

Orders

Hachette UK Distribution, Hely Hutchinson Centre, Milton Road, Didcot, Oxfordshire,
OX11 7HH

tel: 01235 827827

e-mail: education@hachette.co.uk

Lines are open 9.00 a.m.–5.00 p.m., Monday to Friday. You can also order through the
Hodder Education website:
www.hoddereducation.co.uk

© Ferguson Cosgrove 2016

ISBN 978-1-4718-6392-9

First printed 2016

Impression number 6

Year 2022

This guide has been written specifically to support students preparing for the CCEA AS and
A-level Physics examinations. The content has been neither approved nor endorsed by CCEA
and remains the sole responsibility of the author.

Cover photo: kasiastock/Fotolia

Typeset by Integra Software Services Pvt. Ltd, Pondicherry, India

Printed and bound by CPI Group (UK) Ltd, Croydon, CR0 4YY

Hachette UK's policy is to use papers that are natural, renewable and recyclable products and
made from wood grown in well-managed forests and other controlled sources. The logging
and manufacturing processes are expected to conform to the environmental regulations of the
country of origin.

Contents

Getting the most from this book 4

About this book 5

Content Guidance

Physical quantities 6

Scalars and vectors 9

Principle of moments 12

Linear motion 16

Dynamics 25

Newton's laws of motion 27

Linear momentum and impulse 31

Work done, potential and kinetic energy 39

Electric current, charge, potential difference
and electromotive force 44

Resistance and resistivity 47

Internal resistance and electromotive force 59

Potential divider circuits 61

Questions & Answers

Self-assessment test 1 68

Self-assessment test 2 82

Knowledge check answers 94

Index 95

■ Getting the most from this book

Exam tips

Advice on key points in the text to help you learn and recall content, avoid pitfalls, and polish your exam technique in order to boost your grade.

Knowledge check

Rapid-fire questions throughout the Content Guidance section to check your understanding.

Knowledge check answers

1 Turn to the back of the book for the Knowledge check answers.

Summaries

■ Each core topic is rounded off by a bullet-list summary for quick-check reference of what you need to know.

Commentary on sample student answers

Read the comments (preceded by the icon **e**) showing how many marks each answer would be awarded in the exam and exactly where marks are gained or lost.

Commentary on the questions

Tips on what you need to do to gain full marks, indicated by the icon **e**

Sample student answers

Practise the questions, then look at the student answers that follow.

Exam-style questions

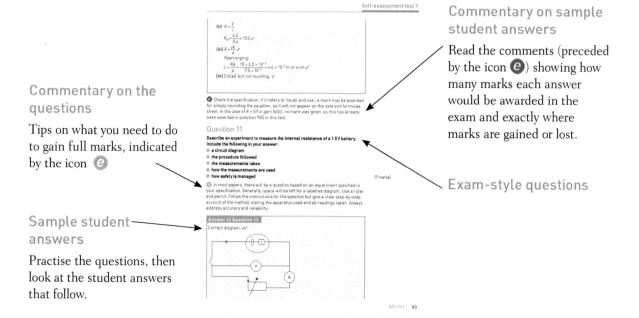

■ About this book

This book covers the CCEA specification for GCE Physics Unit AS 1: Forces, Energy and Electricity. It has two sections:

- The **Content Guidance** covers Unit AS 1. It does not have the detail of a textbook but it offers guidance on the main areas of the content and includes worked examples. These examples illustrate the types of question that you are likely to come across in the examination. The tips will help you to understand the physics and give you guidance on the core aspects of the subject. They also show how to approach revision and improve your exam technique.
- The **Questions & Answers** section comprises two self-assessment tests. Answers are provided and there are comments on the specific points for which marks are awarded.

The specification

The CCEA specification can be obtained from the CCEA website at www.rewardinglearning.org.uk. It describes the physics that is required for the unit assessments, and the format of the assessments.

Symbols, signs and abbreviations

Questions will be set using SI units. You must be familiar with the symbols for quantities relevant to the specification and with the units for these quantities. Questions will assume knowledge of the symbols for the decimal multiples and sub-multiples shown in Table 3 on p. 7.

Content Guidance

■ Physical quantities

Physics is often referred to as the quantitative science. When we measure an amount of any physical quantity, we do not do so in absolute terms, but we compare its size with internationally accepted amounts of a few basic quantities.

All the quantities have magnitude, but what differentiates one quantity from another is its unit.

- A mass of 5 kg can be thought of as 5×1 kg.
- A length of 5 m can be thought of as 5×1 m.

Both examples have the same magnitude but their difference is conveyed by their specific units.

SI base units

It is essential to have a coherent, internationally accepted system of units.

The International System of Units (SI units) has seven **base units**, each exclusive to a fundamental physical quantity (Table 1).

Table 1 Base quantities and SI base units

Base quantity	Name of SI base unit	Symbol for unit
Length	metre	m
Mass	kilogram	kg
Time	second	s
Electric current	ampere	A
Thermodynamic temperature	kelvin	K
Luminous intensity*	candela	cd
Amount of substance	mole	mol

* Not required for the CCEA specification.

Knowledge check 1

What is a base unit?

Knowledge check 2

State six SI base units.

SI derived units

Units for all other physical quantities are **derived units**, obtained from the base units by multiplying or dividing one base unit by one or more other units. For example:

$$\text{speed} = \frac{\text{distance}}{\text{time}} \qquad \text{units: } \frac{\text{metres}}{\text{seconds}}, \text{m s}^{-1}$$

$$\text{charge} = \text{current} \times \text{time} \qquad \text{units: amperes} \times \text{seconds}, \text{A s}$$

Many derived units have their own special name. Those used in the specification are shown in Table 2.

Table 2 Derived quantities and derived units

Derived quantity	Derived unit (in terms of base units)	Name of derived unit	Symbol for unit
Frequency	s^{-1}	hertz	Hz
Force	$kg\,m\,s^{-2}$	newton	N
Pressure	$kg\,m^{-1}\,s^{-2}$	pascal	Pa
Energy	$kg\,m^2\,s^{-2}$	joule	J
Power	$kg\,m^2\,s^{-3}$	watt	W
Charge	As	coulomb	C
Electric potential difference	$kg\,m^2\,s^{-3}\,A^{-1}$	volt	V
Resistance	$kg\,m^2\,s^{-3}\,A^{-2}$	ohm	Ω

Quantities that do not have a named derived unit are expressed in terms of the base unit combination. For example, velocity is measured in metres per second ($m\,s^{-1}$).

Prefixes for SI units

Sub-multiples and multiples of both base and derived units exist. This is to cater for the very large and the very small values that exist with these quantities. There are approved prefixes and symbols (Table 3).

Table 3 Prefixes

Sub-multiple	Prefix	Symbol	Multiple	Prefix	Symbol
10^{-2}	centi	c	10^{3}	kilo	k
10^{-3}	milli	m	10^{6}	mega	M
10^{-6}	micro	μ	10^{9}	giga	G
10^{-9}	nano	n	10^{12}	tera	T
10^{-12}	pico	p			
10^{-15}	femto	f			

The symbol for kilo is an anomaly — unlike other multiples it is *not* a capital; this is because K is used for kelvin.

Equations and units

One of the conditions for an equation to be valid is that each of the terms in the equation must have the same base units. The equation is then said to be **homogeneous**. For example, in

$$v = u + at$$

both v and u are velocities with base units $m\,s^{-1}$. The product at (acceleration multiplied by time) has base units of $m\,s^{-2} \times s$, which also gives $m\,s^{-1}$. Therefore the equation is homogeneous or balanced.

This technique can be used to find the units of an unknown quantity in an equation and hence give a clue to the quantity.

Exam tip

The special names given to many derived units are the names of famous scientists. However, when used for the SI unit, a lower-case first letter must be used, for example hertz. The symbols in these cases do use a capital first letter, for example Hz.

You should never add an 's' to the symbol for a unit to denote the plural, for example 5 Hz, *not* 5 Hzs.

Exam tip

When writing a derived unit, a space is left between the units for each quantity. For example, this distinguishes $m\,s^{-1}$ (which is metres per second) from ms^{-1} (which represents the inverse of milliseconds).

Exam tip

When recording a large or small answer, it is best to present it in standard form, that is, as a number between 1 and 10, multiplied by a power.

Exam tip

Homogeneity shows that an equation *could* be correct, but does not *prove* that it is correct.

Content Guidance

Worked example 1

The unit of stress is the pascal. Express the pascal in base units.

Answer

The pascal, Pa, is the SI unit of pressure.

Recall equations for the quantity for which the base units are known:

$$\text{pressure} = \frac{\text{force}}{\text{area}}$$

$$\text{force} = \text{mass} \times \text{acceleration}$$

Identify the base units for these quantities:

mass: kg

area: m^2

acceleration: $m\,s^{-2}$

Combine these units to find the base units for pressure:

$$\text{base units of pressure} = \frac{(\text{kg} \times m\,s^{-2})}{m^2} = kg\,m^{-1}\,s^{-2}$$

Worked example 2

The equation for Newton's law of gravitation is:

$$F = \frac{GMm}{r^2}$$

Find the base units of G.

Answer

Identify the base units:

base unit of force $F = kg\,m\,s^{-2}$

base unit of masses M and m = kg

base unit of distance r = m

Change the subject of the equation to G:

$$G = \frac{Fr^2}{Mm}$$

Substitute in the base units for each term:

base unit of $G = (kg\,m\,s^{-2})(m^2)/(kg)(kg)$

Simplify by combining and cancelling as you would with numbers:

base unit of $G = kg^{-1}\,m^3\,s^{-2}$

Knowledge check 3

Express 24 nm in standard form.

Exam tip

You should know how to multiply and divide powers. To multiply, add the powers:

$k^x \times k^y = k^{(x+y)}$

To divide, subtract the powers:

$k^x \div k^y = k^{(x-y)}$

Summary

- Physical quantities consist of a numerical magnitude and a unit.
- It is the units that distinguish one quantity from another.
- SI refers to the International System of Units. It is a coherent system based on seven base units.
- The seven base units (and their associated quantities) are: metre (length), kilogram (mass), second (time), ampere (electric current), kelvin (temperature), mole (amount of substance) and candela (luminous intensity).
- Other physical quantities required can be measured in derived units; these are formed by combining base units.
- For an equation to be balanced, the terms on each side of the equation must have the same base units.

Scalars and vectors

Physical quantities can be classified as either **scalar** or **vector** (Table 4).

Table 4 Scalars and vectors

Scalars	Vectors
Mass	Weight
Temperature	Force
Energy	Momentum
Power	Acceleration
Volume	Moment of a force
Time	Displacement
Distance	Velocity
Speed	

Scalar quantities can be fully described by their magnitude and a unit. The direction has no significance — for example, human body temperature is fully described as 37°C and the mass of a person could be 75 kg.

Vector quantities are *not* fully described by *only* their magnitude and a unit — the situation is direction-dependent. A force of 10 N acting on an object will produce a different outcome if it acts to the right compared with if it acted to the left. A vector quantity is only fully described when the direction, magnitude and unit are stated.

Adding scalars and vectors

The addition of scalar quantities is non-problematic, it is a simple arithmetic sum. For example, the total mass of 2 kg plus 3 kg is 5 kg. The increase in temperature from −5°C to 12°C is 17°C.

However, when two vectors are added, we need to take account of their direction as well as their magnitude. There are two methods — by scale drawing or, when they are perpendicular to one another, by use of Pythagoras' theorem.

The sum of two vectors is known as the **resultant**.

Exam tip

Temperature and charge are scalar quantities, yet can have negative values. In these instances, the negative denotes differences in size and type, respectively, and not direction.

Knowledge check 4

State whether the following quantities are scalar or vector: work, density, charge, momentum.

Knowledge check 5

For a falling body, state three scalar and three vector quantities associated with its motion.

Scale drawing

The length of the line, drawn to scale, represents the magnitude and it can be drawn in the direction of the vector. The start of the second vector is drawn from the tip of the first. This is known as the 'nose-to-tail' method.

So to add, or subtract, vectors, you transpose (move in space) the second vector so that you effectively have the action of vector **y** followed by the action of vector **x**. The resultant vector **z** and its direction can be found by direct measurement from the diagram (Figure 1).

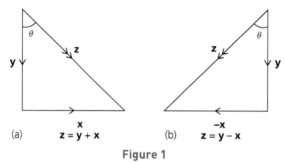

(a)   **z = y + x**   (b)   **z = y − x**

Figure 1

Pythagoras' theorem

To find the magnitude of the resultant, use the equation:

$$x^2 + y^2 = z^2$$

To find the direction of the resultant vector, use trigonometry (where x is the length of vector **x**, and y is the length of vector **y**):

$$\theta = \tan^{-1}\left(\frac{x}{y}\right)$$

Resolution of vectors

Just as we can combine the action of individual vectors to find the single resultant vector, it is sometimes useful to split a vector into its perpendicular components.

Again, the horizontal and vertical components can be found by two methods — scale drawing and trigonometry (Figure 2).

Scale drawing

- The length of the line, drawn to scale, represents the magnitude and it can be drawn in the direction of the vector.
- Lines are drawn horizontally and vertically from the end of the vector.
- Lines are drawn horizontally and vertically from the start of the vector and stop when they meet the previous drawn lines. These lines are the horizontal and vertical components of the vector.

Trigonometry

$\sin\theta = \dfrac{x_v}{x}$ and $\cos\theta = \dfrac{x_h}{x}$

$x_v = x\sin\theta$ and $x_h = x\cos\theta$

Exam tip

Subtraction should be considered as the addition of the negative of the second vector. The negative vector is obtained by reversing the direction of the original vector:

z = y − x or **z = y + (−x)**

Exam tip

Giving the value of θ alone is not sufficient; you must give a compass bearing. For example, in Figure 1 these would be (a) east of south or (b) west of south.

Knowledge check 6

Two forces of magnitude 12N and 5N act on a body. What is:

a the minimum magnitude of the resultant force?
b the maximum magnitude of the resultant force?
c the value of the resultant force when they act at 90° to each other?

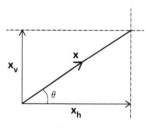

Figure 2

Worked example 1

Two forces of 3 N and 4 N act on a body B, as shown in Figure 3.

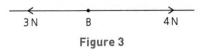

Figure 3

a State the magnitude and direction of the resultant force on B.

b The 4 N force is now turned through 90°, as shown in Figure 4. Find, by calculation, the new resultant force on body B. State the direction of this force relative to the horizontal.

Answer

a resultant force = 4 − 3 = 1 N to the right

b resultant force = $\sqrt{4^2+3^2}$ = 5 N

$\tan \theta = \dfrac{4}{3}$

$\theta = 53°$

The resultant force acts 53° south of west.

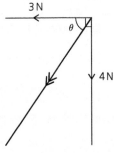

Figure 4

Knowledge check 7

Which is more effective, pushing or pulling a garden roller? Explain your answer.

Knowledge check 8

A tennis player hits a ball with an initial speed of 12.0 m s⁻¹ at an angle of 30° to the horizontal. What are the initial horizontal and vertical components of the speed of the ball?

Worked example 2

A man can row at a speed of 3 m s⁻¹. He wants to row directly across a river. The river current is 2 m s⁻¹. Find the direction in which the man must row and his resultant speed.

Answer

Construct a vector diagram (Figure 5).

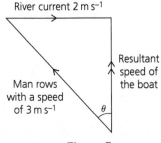

Figure 5

The man must row upstream to allow for the effect of the current:

direction $\theta = \sin^{-1} \dfrac{2}{3} = 42°$

resultant speed of the boat = $\sqrt{3^2-2^2} = \sqrt{5} = 2.2$ m s⁻¹

Summary

- Scalar quantities have magnitude only.
- Vector quantities have magnitude and a direction.
- In a scale drawing, the length of an arrow represents the magnitude of the vector and its direction represents the direction of the vector.
- A vector can be divided or resolved into two perpendicular components to assist in solving problems.

Principle of moments

Rather than cause a body to move in accordance with Newton's law, a different outcome will occur if the object is fixed at a point. The force will cause the object to turn about the point.

Definition of the moment of a force

A familiar example of this concept is seen in the use of a spanner to remove a nut. You apply a force at the end of the spanner to make the nut rotate on the bolt.

By extending the length of the spanner, you can reduce the force needed to release the nut. It is clear that there are two factors that determine the turning effect:

- the perpendicular distance from the pivot
- the size of the force

The turning effect of a force is called the moment of the force. The SI unit of the moment of a force is the newton metre, $N\,m$.

$$\text{moment of force} = F \times d$$

Worked example

A mechanic has to apply a force of $150\,N$ to just loosen a wheel nut on a car. He uses the garage spanner, which is $50\,cm$ long, and applies his force at right-angles at the end of the spanner. What force would the owner of the car have to apply in the same situation if he used his own spanner, which is $20\,cm$ long?

Answer

Calculate the moment the mechanic required:

$$\text{moment of force} = F \times d = 150 \times 0.5 = 75\,N\,m$$

The same moment is required by the car owner:

$$\text{moment of force} = 75 = F \times 0.2$$

$$\text{force applied using short spanner} = F = \frac{75}{0.2} = 375\,N$$

The **moment** of a force about a point is defined as the product of the force and the perpendicular distance of the line of action of the force from the point.

Exam tip

The moment is a vector quantity having directional sense — referred to as clockwise or anticlockwise.

The principle of moments

When a body is acted on by more than one force and is in rotational equilibrium, then the turning effect of the forces must have cancelled each other. The most familiar example of this concept is the see-saw.

A long plank is fixed or pivoted about a centre point and two children sit at either end. The weight of the children provides the force.

From our childhood experiences, we can again confirm that the two factors that determine the size of the turning effect are:

- the distance from the pivot
- the size of the force

In addition, we can intuitively calculate the value of the unknown quantity to achieve a balance. This is based on the principle of moments.

> The **principle of moments** states that, when an object is in rotational equilibrium, the sum of the clockwise moments about a point is equal to the sum of the anticlockwise moments about the same point.

Worked example

Some weights are hung from a rod of negligible mass, as shown in Figure 6. The rod is pivoted as shown. Find the value of the force that must be applied at the end shown to balance the rod.

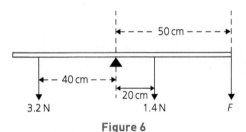

Figure 6

Answer

For the rod in equilibrium:

$$\text{sum of clockwise moments} = (1.4 \times 0.20) + (0.50 \times F)$$

$$\text{sum of anticlockwise moments} = (3.2 \times 0.40)$$

Applying the principle of moments:

$$(1.4 \times 0.20) + (0.50 \times F) = (3.2 \times 0.40)$$

Rearranging:

$$0.50F = (3.2 \times 0.40) - (1.4 \times 0.20)$$

$$0.50F = 1.28 - 0.28$$

Therefore:

$$F = 2.0\,\text{N}$$

Centre of mass/gravity

In the worked example above, we used a light rod. This was so that in the question we could ignore the mass and subsequently the weight of the rod.

The weight of a body when pivoted about a point will produce an unbalanced moment, except when the point is at the centre of mass/gravity. When pivoted about this point, all the contributions to the weight of the body produce moments that cancel: the clockwise and anticlockwise moments are equal. In mathematical terms, it is as if all the mass/weight of the object can be treated as acting through the centre of mass/gravity.

For a uniform body, the centre of mass/gravity will be at the centre of the body. For a non-uniform body, it will be closer to the heavy end, as there is a greater mass contribution from this end.

So in questions on moments, we depict the weight of the body as a single force acting vertically down through the centre of gravity.

Worked example 1

To find the weight of a uniform metre rule, a student set up the arrangement shown in Figure 7. The metre rule was pivoted at the 30 cm mark and a weight of 2.2 N was moved along its length until the ruler was balanced. The weight was at the 19 cm mark when the ruler was balanced.

Use the information given to calculate the weight of the metre rule.

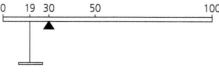

Figure 7

Answer

The body is in rotational equilibrium, so the principle of moments applies:

sum of clockwise moments = sum of anticlockwise moments

The weight of the ruler produces the clockwise moment, and because the body is uniform this can be treated as acting at the 50 cm mark. So:

clockwise moments = $W \times (50 - 30) = 20\,W$

anticlockwise moments = $2.2 \times (30 - 19) = 24.2$

Therefore:

$20\,W = 24.2$

$W = 1.21\,\text{N}$

The weight of the ruler is 1.2 N (2 s.f.).

Worked example 2

A uniform plank AB, 16 m long and weighing 240 N, rests on two supports as shown in Figure 8. The support at X is fixed while the support at Y can be moved. A girl weighing 360 N starts at point X and walks towards B. How far from the right end (B) of the plank should support Y be placed so that she can just reach this end of the plank without causing it to tip?

Figure 8

Answer

When the plank is just about to tip, there will be no contact at support X, and support Y will be the only pivot.

Let Y be a distance s from end B when this happens. Take moments about Y.

The weight of the girl will produce the clockwise moment $360 \times s$.

The weight of the plank will produce the anticlockwise moment $240 \times (8 - s)$.

In rotational equilibrium, the principle of moments applies. Therefore:

$$360 \times s = 240 \times (8 - s)$$

$$360s + 240s = 1920$$

$$600s = 1920$$

$$s = 3.2 \, \text{m}$$

Vertical equilibrium

Where a body is pivoted or supported, and is in rotational equilibrium under the action of a number of forces, Newton's first law will apply, as well as the principle of moments.

The resultant force acting must equal zero: the sum of the upward forces equals the sum of the downward forces.

The upward forces are supplied by the reaction forces at the pivot or support points.

Worked example

A uniform beam, of weight 240 N and length 9.6 m, is supported on two pillars, X and Y, 6.0 m apart. The beam projects 2.0 m from pillar X, as shown in Figure 9. Calculate the support force of each pillar on the beam.

Exam tip

In these calculations, you can always eliminate the turning effect of an unknown force by taking moments about the point through which it acts.

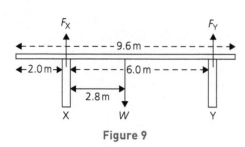

Figure 9

Answer

The structure is in rotational equilibrium, so the principle of moments applies.

Take moments about pillar X:

- Support force F_X produces no turning effect about X.
- Beam weight W produces a clockwise moment.
- Support force F_Y produces an anticlockwise moment.

Then:

clockwise moment = anticlockwise moment

$$(240 \times 2.8) = (F_Y \times 6.0)$$

Therefore:

$$F_Y = 112\,\text{N}$$

The structure is in vertical equilibrium, so $F_{\text{resultant}} = 0$.

forces up = forces down

$$F_X + F_Y = W = 240\,\text{N}$$

As $F_Y = 112\,\text{N}$:

$$F_X = 240 - 112 = 128\,\text{N}$$

Knowledge check 9

A nut needs a moment of 4 N m to turn it. What force is required when using a spanner of length **a** 10 cm, **b** 20 cm?

Knowledge check 10

A crowbar 2.0 m long is levered about a point 0.20 m from its tip. What is the smallest force that must be applied to raise a load of 80 kg at the tip?

Summary

- The moment of a force about a point is defined as the product of the force and the perpendicular distance of the line of action of the force from the point.
- The principle of moments states that, for a body in rotational equilibrium, the sum of the clockwise moments about a point is equal to the sum of the anticlockwise moments about the same point.
- In addition, for complete equilibrium, the resultant force in any direction must be zero.
- The centre of mass/gravity of a body is the point at which all the mass/weight can be treated as acting.

Linear motion

We are all familiar with journeys of one form or another and the quantities involved in describing them:

- How far away is it?
- How long will it take?

- Can't we go any faster?
- Are we there yet?

The quantities involved in describing motion are listed in Table 5.

Table 5 Quantities involved when describing motion

Quantity	Symbol	Unit	Vector or scalar
Distance	s	metres (m)	scalar
Initial speed	u	metres per second (m s⁻¹)	scalar
Final speed	v	metres per second (m s⁻¹)	scalar
Average speed	$\langle v \rangle$	metres per second (m s⁻¹)	scalar
Time	t	seconds (s)	scalar
Displacement	s	metres (m)	vector
Initial velocity	u	metres per second (m s⁻¹)	vector
Final velocity	v	metres per second (m s⁻¹)	vector
Average velocity	$\langle v \rangle$	metres per second (m s⁻¹)	vector
Acceleration	a	metres per second² (m s⁻²)	vector

For everyday descriptions, it is normal to refer to the scalar quantities and a loose interpretation of acceleration. This is fine when we are dealing with motion in a straight line, in one direction, but once a direction change occurs, we must use the vector quantities.

Definitions of quantities

Speed is the rate of change of distance or the distance moved per second:

$$\text{speed} = \frac{\text{distance travelled}}{\text{time taken}}$$

During a journey, the speed may change. Hence we have the concept of average speed for the whole journey:

$$\text{average speed} = \frac{\text{total distance travelled}}{\text{total time taken}}$$

Velocity, a vector quantity, is defined in terms of displacement:

$$\text{velocity} = \frac{\text{displacement}}{\text{time taken}}$$

$$\text{average velocity} = \frac{\text{total displacement}}{\text{total time taken}}$$

During a journey, when the velocity is changing, either in magnitude (speeding up or slowing down) or in direction only, it is accelerating:

$$\text{acceleration} = \frac{\text{change in velocity}}{\text{time taken for the change}}$$

The need for the vector and scalar distinction is best illustrated with a simple example.

Exam tip

When an object is slowing down, acceleration will have a negative value. This is known as deceleration or retardation.

Worked example

A car completes a half lap of a circular track of circumference 6000 m in 150 s (Figure 10). Assume that the car travels at a constant speed.

a What is the constant speed?

b What is the change in velocity after half a lap?

c What is the acceleration during the half lap?

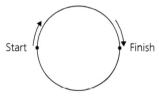

Figure 10

Answer

a $\text{speed} = \dfrac{\text{distance travelled}}{\text{time taken}}$

$\text{speed} = \dfrac{3000}{150} = 20\,\text{m s}^{-1}$

b Note that the car is travelling in *opposite* directions at the start and finish. We use plus and minus signs to denote this difference. So

$$\text{initial velocity } u = +20\,\text{m s}^{-1}$$

$$\text{final velocity } v = -20\,\text{m s}^{-1}$$

$$\text{change in velocity} = v - u$$

$$= (-20) - (20)$$

$$= -40\,\text{m s}^{-1}$$

c $\text{acceleration} = \dfrac{\text{change in velocity}}{\text{time taken}}$

$\text{acceleration} = \dfrac{40}{150}$

$$= 0.27\,\text{m s}^{-2}$$

So, in this example, even though there has been no change of speed, there has been a small acceleration.

Linear motion is motion in a straight line, so the only direction change we will deal with will be up and down or right and left. We can use + and − signs to indicate opposite directions.

Equations of motion

These equations are limited to motion of an object with **constant acceleration in a straight line**. They are based on the basic definitions already given above and one additional point.

For an object moving with constant acceleration, the average velocity $\langle v \rangle$ will be halfway between the initial and the final velocities:

$$\langle v \rangle = \tfrac{1}{2}(u + v)$$

Exam tip

At every point on the track, the speed is the same. However, the velocity is continually changing, because the direction is changing.

Exam tip

Which direction you choose as + is up to you, but be consistent throughout your answer.

Knowledge check 11

A car starting from rest accelerates in a straight line at $4\,\text{m s}^{-2}$ for 5 s. How far has the car travelled in this time?

The four equations of motion are sometimes referred to as the 'suvat' equations, because of the terms used:

Equation 1: $v = u + at$ (no s involved)

Equation 2: $s = \frac{1}{2}(u + v)t$ (no a involved)

Equation 3: $s = ut + \frac{1}{2}at^2$ (no v involved)

Equation 4: $v^2 = u^2 + 2as$ (no t involved)

So, if we know three of the variables, by using the correct equation, we can find the required unknown.

Interpreting displacement–time and velocity–time graphs

A useful way of representing a journey is with a displacement–time or velocity–time graph.

As well as effectively giving an overall summary of the nature of the journey, we can also calculate the unknown values from the appropriate graph.

Displacement–time graph

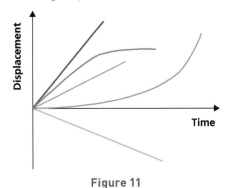

Figure 11

In Figure 11, the coloured lines represent different motions:
- grey — object moving at a constant velocity
- blue — object moving at a greater constant velocity
- orange — object moving at a constant velocity but in the opposite direction
- green — object starting from rest and speeding up
- red — object has a starting velocity and is slowing down to a stop

All this information is based on one principle:

The gradient or slope of a displacement–time graph is equal to the velocity of the object.

It is therefore possible to calculate a value for the velocity from the displacement–time graph.

Knowledge check 12

A car goes from rest to 60 mph in 4.5 s. Calculate its acceleration.

[1 mile = 1.6 km]

Knowledge check 13

A body travelling at 10 m s^{-1} decelerates at 4 m s^{-2}. How far does it travel before coming to a stop?

Knowledge check 14

A car accelerates from 5 m s^{-1} to 25 m s^{-1} in 8 s. How far does it travel in this time and what is the value of its acceleration?

Exam tip

Any journey will be a combination of the lines in Figure 11.

Velocity–time graph

The same descriptions of motion from above can be represented on a velocity–time graph (Figure 12). Here the principle is as follows:

The gradient or slope of a velocity–time graph is equal to the acceleration of the object.

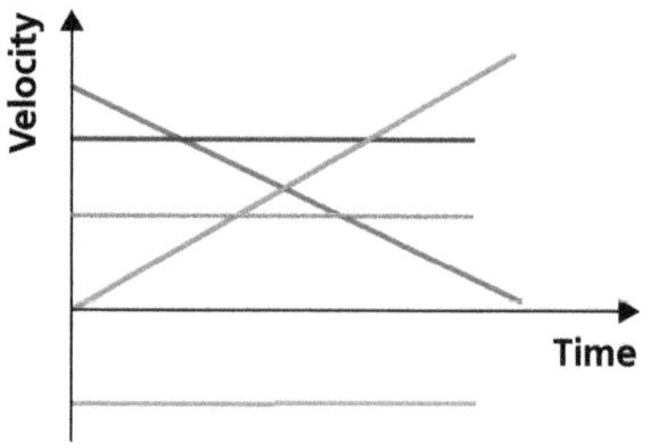

In addition to being able to determine the value of the acceleration from a velocity–time graph, we can evaluate the distance travelled or the total displacement.

Consider the example of a ball thrown vertically upwards, from waist high, and allowed to fall to the ground. The velocity–time graph for the ball is shown in Figure 13. Note the sign convention of up as +.

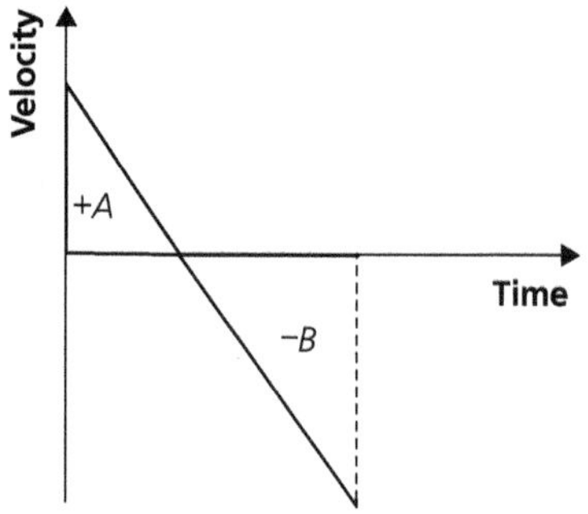

The area between the velocity–time plot and the time axis is equal to the distance travelled by the ball.

So

distance travelled = A + B

If we consider the sign associated with the value of the area, then we get the displacement:

displacement of the ball = A − B

Note that, in our example, area B > area A; the difference is the value from the waist to the ground.

Another possible shape for the velocity–time graph is a curve or non-linear shape (Figure 14). This represents changing acceleration. A classic example of non-uniform acceleration occurs in a parachute jump. The change in the acceleration is a consequence of the change in the force acting on the body.

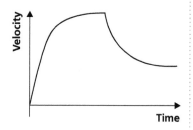

Figure 14

Initially, the parachutist is in free fall, but, as her velocity increases, the upward force of air resistance increases, causing the acceleration to decrease (slope decreasing). When no resultant force acts, the acceleration is zero (horizontal line). The parachute is opened, increasing the frictional force, and the parachutist decelerates (negative slope), until once again the resistive force is equal and opposite to the weight and she falls at a constant 'terminal' velocity (horizontal line).

Figure 12

Figure 13

Exam tip

It is possible to draw distance–time and speed–time graphs, but their use is limited because they cannot convey direction reversal. (Displacement–time and velocity–time graphs can convey direction reversal.)

Experiment to measure the acceleration of free fall

Because of the variable force of attraction of the Earth on objects of different mass, they will all fall with the same uniform acceleration. This acceleration is called the **acceleration of free fall** and is represented by the symbol g.

There are a variety of methods used to determine g.

Use of $s = ut + \frac{1}{2}at^2$

If we ensure that the object falls from rest, i.e. $u = 0$, the equation will reduce to

$$s = \tfrac{1}{2}gt^2$$

In an experiment, we time the fall of an object, from rest, through a range of different distances, noting the time of fall in each case. A graph of s against t^2 will be linear, with a slope equal to a value of $\frac{1}{2}g$.

As the time values are so small, it is usual to use an electrical timer, which is triggered to commence timing as a steel ball is released, and to stop timing as the ball breaks a circuit by opening a trap door (Figure 15).

Use of $a = (v - u)/t$

Here we use a card of known length to fall through two light gates attached to a datalogger and computer. The software will measure time values to one-hundredth of a second. Three times are automatically recorded:

- the time for the card to fall through gate A, t_A
- the time for the card to fall through gate B, t_B
- the time for the card to fall from gate A to gate B, t_{AB}

The length, s, of the card is measured and entered into the computer. Then

initial velocity $\qquad u = \dfrac{s}{t_A}$

final velocity $\qquad v = \dfrac{s}{t_B}$

acceleration of free fall $g = \dfrac{v - u}{t_{AB}}$

With the datalogger and light gate system, it is possible to obtain numerous results very quickly, without the need for graphical analysis. This allows the opportunity for numerous repeats and variation of factors — for example, changing the mass of the card.

Worked example 1

A jet aircraft accelerates from rest for 30.0 s, leaving the ground at a speed of 285 kilometres per hour.

a Find the take-off speed, in metres per second.

b Calculate the average acceleration of the aircraft during take-off.

c Calculate the length of runway used by the aircraft during take-off.

→

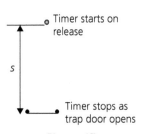

Figure 15

Answer

a Take-off speed:

$$285\,km\,h^{-1} = 285 \times 1000\,m\,h^{-1}$$

$$= \frac{285\,000}{60 \times 60}\,m\,s^{-1} = 79.2\,m\,s^{-1}$$

b Average acceleration:

$$a = \frac{v - u}{t}$$

$$a = \frac{79.2 - 0}{30} = 2.64\,m\,s^{-2}$$

c Displacement:

$$s = \tfrac{1}{2}(u + v)t$$

$$s = \tfrac{1}{2}(0 + 79.2) \times 30 = 1188\,m$$

Worked example 2

A cricketer throws a ball vertically upwards with a velocity of $14.0\,m\,s^{-1}$ and catches it as it falls.

a How high does the ball go?

b How long does it take to reach this height?

c How long is the ball in the air?

Answer

a Consider the flight of the ball up to the highest point.

At the highest point, the ball is momentarily at rest. Therefore:

velocity at highest point, $v = 0$

initial velocity, $u = 14.0\,m\,s^{-1}$, upwards

The acceleration is constant throughout and in the opposite direction from the initial velocity. Therefore:

$$a = g = -9.81\,m\,s^{-2}$$

Use the equation:

$$v^2 = u^2 + 2as$$

Rearranging:

$$s = \frac{v^2 - u^2}{2a}$$

$$s = \frac{0 - 14.0^2}{2(-9.81)}$$

$$s = \frac{-196}{-19.62} = +9.99\,m$$

➡

b Use the equation:

$$v = u + at$$

Rearranging:

$$t = \frac{v - u}{a}$$

$$t = \frac{0 - 14.0}{-9.81} = 1.43\,\text{s}$$

c As the ball is caught at the same point from which it was propelled, the overall displacement is zero. This allows direct solution using the equation:

$$s = ut + \tfrac{1}{2}at^2$$

$$0 = 14.0t + \tfrac{1}{2}(-9.81)t^2$$

Rearranging:

$$14.0t = 4.9t^2$$

$$t = \frac{14.0}{4.9} = 2.86\,\text{s}$$

> **Exam tip**
>
> Note that the answer to part **c** is twice the answer to part **b**. This is no surprise, as the motion has symmetry. But this will only be true when the net vertical displacement is zero.

Worked example 3

A ball is dropped from rest onto the ground and is allowed to bounce before being caught at the highest point. The velocity–time graph for the motion of the ball is shown in Figure 16.

> **Exam tip**
>
> When the ball is falling, it has negative velocity. When the ball is rising, it has positive velocity.

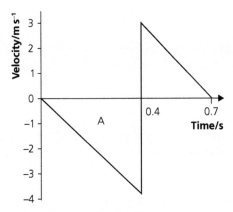

Figure 16

a i   With what velocity does the ball hit the ground?
   ii With what velocity does it leave the ground?
b Use the graph to find the acceleration of free fall.
c From what height was the ball released?

Answer

a At 0.4 s the velocity of the ball suddenly changes from a negative to a positive value. This indicates a change of direction, which occurs when the ball hits the ground.

   i   The velocity as it hits the ground is $-4.0\,\mathrm{m\,s^{-1}}$ (read directly from the graph).

   ii The velocity as it leaves the ground is $+3.0\,\mathrm{m\,s^{-1}}$ (read directly from the graph).

b The acceleration is equal to the slope of the velocity–time graph. Two slopes can be determined from the graph.

As it is falling:
$$a = g = \frac{\Delta v}{t} = \frac{-4 - 0}{0.4} = -10\,\mathrm{m\,s^{-2}}$$

As it is rising:
$$a = g = \frac{\Delta v}{t} = \frac{0 - 3.0}{0.3} = -10\,\mathrm{m\,s^{-2}}$$

Average value of acceleration, $g = -10\,\mathrm{m\,s^{-2}}$

(The negative indicates a downward direction.)

c The distance travelled is equal to the area between the velocity–time graph and the time axis.

We want the distance travelled from when the ball is released and when it hit the ground, i.e. from $t = 0$ to $t = 0.4\,\mathrm{s}$.

We want the area of triangle A, labelled in the diagram:

$$\text{area of triangle A} = \tfrac{1}{2} \times \text{base} \times \text{perpendicular height}$$

$$= \tfrac{1}{2} \times 0.4 \times 4.0 = 0.8\,\mathrm{m}$$

Summary

- Displacement, a vector quantity, is defined as the change in position from a starting or reference point.
- Speed is defined as the distance moved per second.
- Velocity is defined as the rate of change of displacement.
- Acceleration is defined as the rate of change of velocity.
- For linear motion problems, we can use + and – to indicate opposite directions.
- For bodies moving with constant acceleration in a straight line:

$$v = u + at$$
$$s = \tfrac{1}{2}(u + v)t$$

$$s = ut + \tfrac{1}{2}at^2$$
$$v^2 = u^2 + 2as$$

- The gradient of a displacement–time graph is equal to the velocity of a body.
- The gradient of a velocity–time graph is equal to the acceleration of a body.
- The area between a velocity–time plot and the time axis gives us the information to determine distance travelled and the displacement of the body.
- In the absence of air resistance, a body falling freely near to the Earth's surface will experience the same acceleration, independent of its mass. The acceleration, g, has a value of $9.81\,\mathrm{m\,s^{-2}}$.

■ Dynamics

Description of projectile motion

So far we have considered the motion of objects in a straight line, but another common situation exists — projectile motion.

Any object propelled at an angle to the vertical near the Earth's surface is an example of projectile motion.

A charged particle moving at an angle to a uniform electric field is another example of projectile motion.

In projectile motion the particles will follow a curved path known as a **parabola**.

Explanation of projectile motion

Consider the most common case of a thrown particle. It is subject to a vertical gravitational force — its weight. The horizontal force of air resistance is neglected.

Whilst moving forwards, the particle also moves downwards under the effect of gravity. The resultant path of the projectile is in effect a combination of two motions — horizontal and vertical.

This allows us to work with the horizontal and vertical aspects of motion separately, and, as they are in straight lines, apply the equations of motion.

Note that the time of flight is the same for both.

To find the actual velocity of the particle, we simply add the components vectorially (Figure 17).

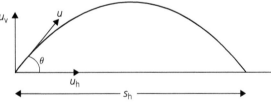

Figure 17

The terms in Table 6 can be applied to all projectile questions.

Table 6 Terms applied to projectile motion

Horizontal component	Vertical component
$u_h = u\cos\theta$	$u_v = u\sin\theta$
$a_h = 0$	$a_v = g$
s_h = range	$s_v = \Delta h$ = change in height
t_h = time of flight	t_v = time of flight

Let us apply this technique to a simple problem.

Projectile motion is motion in two dimensions: the object is acted on by a constant force in one dimension and no force in the other dimension.

Exam tip

The vertical force has no component in the horizontal plane and cannot therefore cause any change to its horizontal velocity component.

Worked example

A stone is thrown horizontally from the top of a cliff 60 m high, with a velocity of 12 m s⁻¹.

a How long does it take to reach the ground below?

b How far from the base of the cliff does it hit the ground?

c What is its speed and direction on hitting the ground?

Answer

Apply this to the general horizontal and vertical terms. Note that for this question we take the downwards direction as positive.

Horizontal component	Vertical component
$u_h = u\cos\theta = 12\cos0 = 12\,\text{m s}^{-1}$	$u_v = u\sin\theta = 12\sin0 = 0$
$a_h = 0$	$a_v = g = 9.81\,\text{m s}^{-2}$
$s_h = \text{range}?$	$s_v = \Delta h = \text{change in height} = 60\,\text{m}$
$t_h = \text{time of flight}$	$t_v = \text{time of flight}$

a Use the vertical component values and $s = ut + \frac{1}{2}at^2$:

$$60 = 0 + \tfrac{1}{2}(9.81)(t^2)$$

$$t^2 = 12.2$$

$$t = 3.5\,\text{s}$$

b Use the horizontal component values and $s = \frac{1}{2}(u + v)t$.

The horizontal component of the velocity does not change, so $u = v$:

$$s_h = \tfrac{1}{2}(12 + 12) \times (3.5) = 42.0\,\text{m}$$

c Acceleration due to gravity means that the vertical component of velocity is increasing. Use the equation $v = u + at$:

$$v_v = u_h + gt$$

$$= 0 + (9.81 \times 3.5) = 34.3\,\text{m s}^{-1}$$

The resultant or actual velocity of the stone is the vector sum of the two components:

$$v_r = \sqrt{v_h{}^2 + v_v{}^2} = \sqrt{12.0^2 + 34.3^2} = 36.4\,\text{m s}^{-1}$$

The direction of the stone can be found by reference to the vector diagram (Figure 18):

$$\tan\theta = \frac{v_v}{v_h} = \frac{34.3}{12.0}$$

$$\theta = 70.7° \text{ below the horizontal}$$

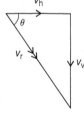

Figure 18

Knowledge check 16

What name is given to the flight path followed by a projectile?

Knowledge check 17

An object is projected with a velocity of 10.0 m s⁻¹ at an angle of 30° to the horizontal. State the values of the acceleration, the vertical component of the velocity and the horizontal component of the velocity **a** at the start, **b** at the top of its flight.

Knowledge check 18

A projectile lands at the same vertical height from which it is launched. It is in the air for 6.8 s. How long did it take to reach its highest point?

Summary

- Projectiles follow a parabolic path.
- A projectile has a uniform velocity in one direction and a uniform acceleration in a perpendicular direction.
- The equations of motion can be applied independently to the two perpendicular components of the projectile motion.

Newton's laws of motion

We have described the motion of objects by graphical means and through the equations of motion. But now we consider what causes objects to move.

- From our everyday experiences, it is obvious that force is required.
- A body requires a force to move it from rest or to bring it to rest.
- A body requires a force to change its speed or direction.

It was Isaac Newton who cleared up the misconception that a force was needed to keep a body moving with a constant velocity. He stated that no **resultant** force acted; rather, a force was needed to overcome unseen frictional, resistive forces.

Newton summarised his theory of motion in three laws. (There are several versions of the wording for these laws.)

1 **A body will remain at rest or continue to move with a constant velocity unless acted on by a resultant force.**

2 **For a body of constant mass, its acceleration is directly proportional to the resultant force applied and in the direction of the resultant force.**

3 **Whenever one body exerts a force on another, the second body exerts an equal and opposite force on the first body.**

Newton's laws are more easily visualised when we consider a near friction-free environment, for example on an ice rink.

1 Once moving, a skater continues to glide with the same speed in a straight line.

2 When a force is applied to the skater, their velocity changes; acceleration takes place.

3 If two stationary skaters, A and B, face one another and skater A pushes skater B, both are seen to move in opposite directions. To the action of A on B there is a reaction of B on A.

Friction: a force that opposes motion

Frictional forces are the unseen forces that need to be considered if everyday situations are to be seen to obey Newton's laws. These frictional forces:

- will always oppose motion
- are reactionary forces — they will only act to oppose action forces
- give rise to the development of heat energy
- often occur when a moving body is in contact with another medium
- can be helpful, for example the grip needed when walking and braking
- can be a hindrance, for example the wear in moving engine parts

Exam tip

Forces are vector quantities; the SI unit of force is the newton (N).

Exam tip

The first and second laws both comply with the equation:

$$F_{resultant} = ma$$

Exam tip

It is important to emphasise in the third law that the acting force and the reacting force act on *different* bodies.

Exam tip

Use of a lubricant indicates that the friction effect is not helpful.

Worked example 1

A body of weight 30.0 N is suspended from a fixed point by a vertical string. Another force pulls the body sideways along the horizontal until the string is at an angle of 35° to the vertical. The system is then in equilibrium.

a Sketch a diagram to show all the forces acting on the object. Clearly label and name the forces and indicate their direction.

b Calculate the magnitude of the tension in the string.

c Find the value of the horizontal pulling force.

Answer

a

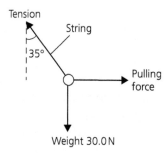

Figure 19

b The body is at rest, so the sum of the components in the horizontal and vertical planes must be zero.

Consider the vertical components:

$$F_{resultant} = 0 = T\cos 35° - W$$

Therefore:

$$T\cos 35° = 30.0$$

$$T = 36.6\,\text{N}$$

c Consider the horizontal components:

$$F_{resultant} = 0 = T\sin 35° - F_{pull}$$

Therefore:

$$F_{pull} = (36.6\sin 35°) = 21.0\,\text{N}$$

Worked example 2

A pushing force of 5.0 N is applied to a body of mass 2.5 kg. The body accelerates at 1.8 m s^{-2}. Find the frictional force that is acting on the body.

Answer

It is the resultant force that causes the acceleration:

$$F_{resultant} = F_{push} - F_{friction} = ma$$

$$F_{resultant} = 5.0 - F_{friction} = (2.5 \times 1.8)$$

$$F_{friction} = 5.0 - 4.5 = 0.5\,\text{N}$$

Weight and Newton's second law

Any body close to the surface of the Earth will fall to Earth with the same acceleration g, known as the acceleration due to gravity or the acceleration of free fall.

It was Galileo who first confirmed that g was independent of mass. For an object of larger mass to fall with the same acceleration as a lighter mass, a greater force must act.

According to Newton's second law:

$$F = mg$$

This force with which an object is pulled towards the Earth is called its **weight**, W. So $W = mg$ is one specific example of Newton's second law.

Experimental verification of Newton's laws

The arrangement shown in Figure 20, with a trolley pulled along a runway by the force provided by a connected falling mass, can be used to investigate Newton's first and second laws of motion.

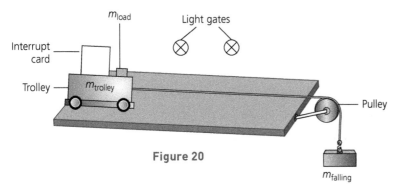

Figure 20

The datalogger and light gate system, used in the experiment to measure the acceleration of free fall, are again used to measure acceleration.

Hanging slotted masses provide a steady known tension in a connecting string, which pulls the trolley along the runway.

Verification that $a \propto F$ for a fixed mass

The tricky part of this investigation is keeping the mass that is moved by the force constant, while changing the mass that is falling, as it provides the force. We have

$$F_{resultant} = m_{moving}a$$
$$m_{moving} = m_{trolley} + m_{load} + m_{falling}$$
$$F = \text{weight} = m_{falling}a$$

The force can be changed without altering the overall mass of the system by placing any masses from $m_{falling}$ onto the back of the trolley, m_{load}.

Start with F at its maximum value, e.g. 8 N (therefore $m_{falling} \approx 800$ g, 7 × 100 g slotted masses and a mass hanger) and no mass on the back of the trolley.

Release the trolley and record the acceleration. Repeat this until three consistent values for the acceleration are obtained.

Knowledge check 19

The total mass of a lift and its passengers is 1000 kg. The lift ascends at a constant speed of 1 m s^{-1}. What is the tension in the cable pulling the lift?

Knowledge check 20

A car of mass 1200 kg accelerates uniformly from rest to a velocity of 12 m s^{-1} in 24 s. Calculate the force that produced this acceleration.

Knowledge check 21

The tension in the towrope pulling a water skier of mass 80 kg is 560 N. What is the value of the frictional force acting on the skier if his acceleration is 1.5 m s^{-2}?

Then remove 100 g from the hanging mass and place it onto the back of the trolley. Once again, release the trolley and record the acceleration. Repeat the procedure until 700 g are on the back of the trolley and only 100 g, the mass hanger, is providing a 1 N force to accelerate the loaded trolley along the runway.

Plot a graph of force against acceleration.

The expected outcome is a straight line *but not through the origin*.

Mapping the quantities against the general straight-line equation may help to explain things:

$$y = mx + c$$
$$F = m_{moving}a + ?$$

We can identify the gradient as equal to the fixed value of m_{moving}.

This can be checked by measuring the mass of the trolley used and adding the maximum falling mass, 800 g.

But what does the intercept on the y-axis represent?

It is the force required to cause zero acceleration — in effect, the force needed to overcome any frictional forces, $F_{friction}$:

$$F_{resultant} = W - F_{friction} = m_{moving}a$$

Typical results for the experiment are shown in Table 7.

1 Use the results to plot a graph of force against acceleration.

2 Use the graph to show that the mass of the trolley is about 1 kg.

3 Determine the frictional force acting on the trolley.

Table 7 Typical experimental results for a fixed moving mass*

$m_{falling}$/g	800	700	600	500	400	300	200	100
a/m s^{-2}	4.13	3.50	2.99	2.55	1.99	1.45	0.91	0.37
	4.10	3.56	3.11	2.55	1.98	1.45	0.90	0.38

*Anomalous results have been removed.

Verification of $a \propto 1/m$, when force acting is constant

In this experiment, the same apparatus is used. However, $m_{falling}$ remains constant and m_{moving} is altered by adding masses onto the back of the trolley, that is, m_{load} is changed.

Typical results for the experiment are shown in Table 8. The trolley used may be assumed to have a mass of 1 kg and m_{faling} = 400 g.

1 Complete Table 8 and plot a graph of acceleration against $1/m_{moving}$.

2 Use your graph to determine the resultant force acting on the trolley and explain why your answer is slightly less than $F = m_{falling}g = 3.9$ N.

The fact that the graphs produced in these experiments are straight lines and that the values determined from the graphs are in agreement with the actual practical values can be viewed as verification of $F_{resultant} = ma$, Newton's second law. But note that it was assumed in this experiment that $W = mg$.

Exam tip

An alternative strategy is to tilt the runway so that you 'friction-compensate'. The angle needed is checked by giving the trolley a slight initial push, allowing it to pass through the two light gates with no pulling force. Adjust the runway angle until an acceleration of zero is recorded. The trolley is moving with a constant velocity, and Newton's first law demonstrated.

Table 8 Typical experimental results for a fixed force*

m_{load}/g	0	200	400	600	800
a/m s^{-2}	2.58	0.63	0.56	0.50	0.46
	2.50	0.62	0.55	0.50	0.45
m_{moving}/kg					
$(1/m_{moving})$/kg^{-1}					

*Remember, $m_{trolley}$ = 1000 g or 1 kg and $m_{moving} = m_{trolley} + m_{load} + m_{falling}$.

Summary

- Newton's first law states: a body will remain at rest or continue to move with a constant velocity unless acted on by a resultant force.
- Newton's second law states: for a body of constant mass, its acceleration is directly proportional to the resultant force applied and in the direction of the resultant force.
- Newton's third law states: whenever one body exerts a force on another, the second body exerts an equal and opposite force on the first body.
- Friction is a force that opposes motion.
- Weight is the force with which a body is pulled towards the Earth's surface and is given by $W = mg$. This is an application of Newton's second law.

■ Linear momentum and impulse

Momentum

All moving bodies have momentum. The concept of momentum is used by sports people in considering their 'follow through', and by car designers in the safety design features.

Momentum is a vector quantity with its direction being that of the body's velocity.

The momentum of a body is defined as the product of its mass and its velocity:

momentum = mass × velocity

$$p = mv$$

The base units of momentum are kg m s^{-1}.

Newton's second law of motion and momentum

Newton was conscious of the concept of momentum, and his second law was originally expressed in terms of momentum:

The change of momentum per second is equal to the applied force, and the momentum change takes place in the direction of the force.

Exam tip

The momentum of a body will remain constant unless an external force acts on it.

Exam tip

It is important to remember that momentum is a vector quantity and, as such, has both a magnitude and a direction. This becomes very important when moving bodies collide and changes in momentum take place.

Thus:

$$\text{momentum change} = mv - mu = \Delta p$$
$$F = \frac{(mv - mu)}{t} = \frac{\Delta p}{t}$$
$$F = \frac{m(v - u)}{t}$$

But

$$\frac{(v - u)}{t} = a$$

Therefore

$$F = ma$$

This is the form of Newton's second law with which you are more familiar.

Impulse

We have just used the equation:

$$F = \frac{\Delta p}{t}$$
$$Ft = \Delta p$$

The product force × time is known as the **impulse** of the force on a body, and the time is that for which the force acts on the body.

impulse = change in momentum = $mv - mu$

Exam tip

The force referred to is the resultant of any forces acting on the body.

Worked example

A rugby player of mass 80 kg is moving forwards with a speed of 3.5 m s^{-1}. He is tackled and moved back in the opposite direction with the same speed.

a Calculate the momentum of the player before and after the tackle.

b Calculate the impulse of the force exerted on the player in the tackle.

c If the tackle lasts for 1.4 s, what is the average force exerted on the player in the tackle?

Answer

a Before: $p = mu = 80 \times 3.5 = 280 \, \text{kg m s}^{-1}$
   After: $p = mv = 80 \times (-3.5) = -280 \, \text{kg m s}^{-1}$

b Impulse = $mv - mu = -280 - 280 = -560 \, \text{kg m s}^{-1}$.

c Impulse = $F \times t$
   Average resultant force is:

$$F = \frac{\text{impulse}}{\text{time}} = \frac{560}{1.4} = 400 \, \text{N}$$

Note that, if the vector nature of momentum had not been recognised in this example, we would be suggesting no momentum change and therefore no force in the tackle.

Knowledge check 23

Show that the unit of impulse, N s, and the unit of momentum, kg m s^{-1}, are equivalent.

Exam tip

Momentum and velocity are vectors. If you assign positive values to one direction then the opposite direction must be assigned a negative value.

Momentum and car safety

In most substantial collisions, the front of a car stops almost immediately. However, the passengers do not. They obey Newton's first law, and continue moving forwards with the same momentum, until a force changes their motion. If they are not restrained, this force will be provided by an impact with the front seat, another passenger, the steering wheel or the windscreen.

The seat belt serves two purposes: it prevents the passenger coming into contact with these other objects, and it is designed to provide a restraining force over a longer period of time by stretching about 25 cm.

Crumple zone technology is based on the same principle. Crumple zones are built into car bonnets, boots and bumpers. By crumpling, the car takes a longer time to come to rest. This again means a lower rate of change of momentum.

The momentum change in the collision cannot be altered, but, by increasing the time over which the change occurs, the forces acting on the passengers and driver are reduced.

Momentum and the 'follow through'

Why do you 'follow through' in sports such as tennis and cricket? By following through, the force imparted to the ball acts for a longer time. The impulse given to the ball is increased, and in consequence there is a greater momentum change. In essence, the ball leaves the racquet, bat, stick, club or boot with a greater speed.

Worked example 1

A person of mass 80 kg jumps from a height of 5 m, landing on the ground with a velocity of 10 m s^{-1}. He does not flex his knees in the landing and is brought to rest very quickly, in 0.1 s.

a Calculate the force acting on him when he lands.

If he had flexed his knees on landing, then he would have been brought to rest much more slowly.

b i Calculate the force acting on him if he flexes his knees and is brought to rest in 0.5 s.

ii Explain the benefit of flexing the knees on landing.

Answer

a $F = \dfrac{\Delta p}{t} = \dfrac{(80 \times 0) - (80 \times 10)}{0.1} = -8000\,\text{N}$

b i $F = \dfrac{\Delta p}{t} = \dfrac{(80 \times 0) - (80 \times 10)}{0.5} = -1600\,\text{N}$

ii There is less impact force, so less damage is done to the body on landing.

Worked example 2

A golfer hits a ball of mass 45 g at a speed of 38 m s^{-1}. The average force exerted by the club head on the ball when in contact is 580 N. Calculate the time the club head and ball are in contact.

Answer

impulse = change in momentum

$$F \times t = mv - mu$$

and so:

$$t = \frac{m(v - u)}{F}$$

$$t = \frac{45 \times 10^{-3} \times (38 - 0)}{580}$$

$t = 2.95 \times 10^{-3}$ s $= 2.95$ ms (3 ms to 2 significant figures)

Impulse and force–time graphs

The force applied to an object when it is hit is most likely to change during the impact. Figure 21 shows how the force F acting on a tennis ball during a serve changes with time t.

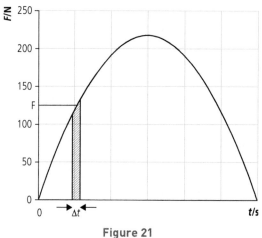

Figure 21

The area of the shaded strip $= F \times \Delta t =$ the impulse added during a small time interval Δt.

Repeat this process, dividing the whole area under the curve into strips and finding the impulse for each interval. Adding together the areas would then give the total impulse given to the ball during the serve.

Thus the area between the force–time curve and the time axis gives the total impulse.

Principle of conservation of momentum

Now consider the momentum of both bodies interacting. This can be investigated experimentally in the classroom using the same apparatus as for the confirmation of Newton's laws — the runway with trolleys or linear air track with glider systems, as shown in Figures 22 and 23.

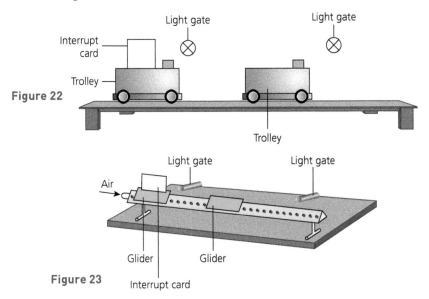

Figure 22

Figure 23

The **principle of the conservation of linear momentum** states that, if no external forces act on a system of interacting bodies, the total momentum of the objects in a given direction before collision equals the total momentum in the same direction after collision.

Conservation of momentum is seen as a fundamental law of physics and applies to all interactions, a traditional collision, explosions, radioactive decays, fission and fusion events.

In equation form, for a two-body system, it can be expressed as:

$$m_1 u_1 + m_2 u_2 = m_1 v_1 + m_2 v_2$$

External forces

No external agent must be acting on the interacting bodies, for example friction, otherwise momentum may be added to or taken from the system.

Sometimes momentum does appear to be gained, for example a body falling towards the Earth. The body will accelerate due to gravity, its velocity will increase and momentum is gained. However, the body is interacting with the Earth, which will experience an equal gain in momentum in the upward direction. As the mass of the Earth is so large, the velocity change of the Earth is negligible.

Momentum conservation and Newton's third law of motion

We have seen the link between Newton's second law and the impulse–momentum equation. Newton's third law can also be shown to be linked, indeed equivalent.

Newton's third law can be expressed as:

$$F_{1 \text{ on } 2} = -F_{2 \text{ on } 1}$$

Using the second law:

$$m_2 a_2 = -m_1 a_1$$

Using the definition of acceleration:

$$\frac{m_2(v_2 - u_2)}{t} = \frac{-m_1(v_1 - u_1)}{t}$$

where t is the interaction time.

Expanding and cancelling t:

$$m_2 v_2 - m_2 u_2 = -m_1 v_1 + m_1 u_1$$

Rearranging:

$$m_1 u_1 + m_2 u_2 = m_1 v_1 + m_2 v_2$$

Application of the conservation of momentum

Consider two classic but contrasting situations and apply the principle of the conservation of momentum.

The first situation is as shown in Figure 24, in which a moving body hits a stationary body and both move off together:

$$u_2 = 0 \qquad \text{and} \qquad v_1 = v_2 = v$$

Then $m_1 u_1 + m_2 u_2 = m_1 v_1 + m_2 v_2$ becomes:

$$m_1 u_1 = (m_1 + m_2)v$$

and if the bodies have equal mass, $m_1 = m_2 = m$:

$$m u_1 = 2mv$$

Therefore $u_1 = 2v$, they move off at half the speed in the same direction.

This situation is seen in railway truck coupling.

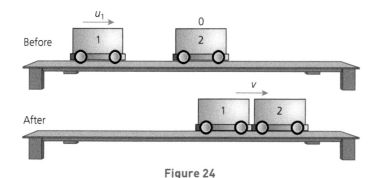

Figure 24

Knowledge check 24

A bowl with a mass of 1.5 kg moves with a speed of 1.2 m s^{-1} on a smooth, horizontal green. It strikes a stationary bowl of mass 1.2 kg head-on. The 1.2 kg bowl begins to move with a speed of 0.9 m s^{-1}. What is the speed of the 1.5 kg bowl immediately after the collision?

Worked example

A railway truck of mass $1200\,\text{kg}$ is rolling along a level track at a speed of $6.0\,\text{m s}^{-1}$ towards a stationary truck. On collision, the trucks become joined and now move with a common velocity of $2.0\,\text{m s}^{-1}$. Find the mass of the second truck.

Answer

First write down what is given:

$$m_1 = 1200\,\text{kg}$$

$$u_1 = 6.0\,\text{m s}^{-1}$$

$$u_2 = 0$$

$$v_1 = v_2 = v = 2.0\,\text{m s}^{-1}$$

Then the equation:

$$m_1 u_1 + m_2 u_2 = m_1 v_1 + m_2 v_2$$

becomes:

$$m_1 u_1 = (m_1 + m_2)v$$

$$1200 \times 6.0 + 0 = (1200 + m_2) \times 2$$

$$3600 = (1200 + m_2)$$

$$m_2 = 2400\,\text{kg}$$

The second situation is the 'explosion' of a stationary body. This situation is simulated by clipping together two identical trolleys so as to compress a spring between them, as shown in Figure 25. On release of the clip, the spring will uncoil and the trolleys separate. In this situation:

$$u_1 = u_2 = 0$$

Then $m_1 u_1 + m_2 u_2 = m_1 v_1 + m_2 v_2$ becomes:

$$0 = m_1 v_1 + m_2 v_2 \qquad \text{or} \qquad m_1 v_1 = -m_2 v_2$$

and if the bodies have equal mass, $m_1 = m_2 = m$:

$$v_1 = -v_2$$

and the bodies move off at the same speed in opposite directions.

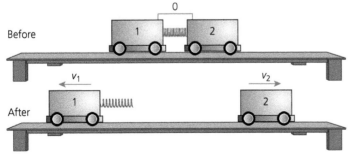

Figure 25

Worked example 1

A bullet of mass 6.0 g is fired from a gun of mass 3.0 kg with a velocity of $200\,\mathrm{m\,s^{-1}}$ (see Figure 26). Find the velocity with which the gun recoils.

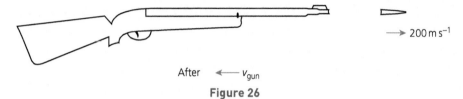

After $\longleftarrow v_{\mathrm{gun}}$

$\longrightarrow 200\,\mathrm{m\,s^{-1}}$

Figure 26

Answer

Both bullet and gun are stationary to start with. Therefore $0 = m_1v_1 + m_2v_2$ becomes:

$$0 = m_{\mathrm{bullet}}v_{\mathrm{bullet}} + m_{\mathrm{gun}}v_{\mathrm{gun}} \quad \text{or} \quad m_{\mathrm{bullet}}v_{\mathrm{bullet}} = -m_{\mathrm{gun}}v_{\mathrm{gun}}$$

$$0 = (6 \times 10^{-3}) \times 200 = -3 \times v_{\mathrm{gun}}$$

$$v_{\mathrm{gun}} = -0.4\,\mathrm{m\,s^{-1}}$$

The gun recoils with a velocity of magnitude $0.4\,\mathrm{m\,s^{-1}}$ into the shoulder.

Worked example 2

A trolley A of mass 0.20 kg and a second trolley B of mass 0.40 kg are clipped together and placed on a smooth horizontal surface (Figure 27). A spring between the trolleys is compressed when they are clipped together. The clip is released and the trolleys move apart. The speed of A is $2.4\,\mathrm{m\,s^{-1}}$.

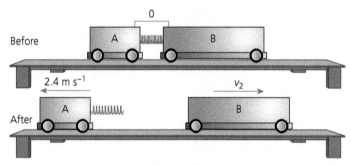

Figure 27

a Calculate the speed of trolley B.
b Calculate a minimum value for the energy stored in the spring when compressed.

Answer

a Apply $0 = m_1v_1 + m_2v_2$ or $m_1v_1 = -m_2v_2$:

$$0.20 \times 2.4 = -(0.40 \times v_2)$$

$$v_2 = -1.2\,\mathrm{m\,s^{-1}}$$

b Use $E_{before} = E_{after}$:

$E_{stored} = KE_{trolleys}$

$E_{stored} = \frac{1}{2}m_A v_A{}^2 + \frac{1}{2}m_B v_B{}^2$

$E_{stored} = (\frac{1}{2} \times 0.2 \times 2.4^2) + (\frac{1}{2} \times 0.4 \times [-1.2]^2)$

$E_{stored} = (0.58) + (0.29) = 0.87\,J$

Note that the direction of the velocity becomes irrelevant when squared and used to determine energy, which is a scalar quantity.

Such an 'explosion' type of interaction is classed as *inelastic* in type as we move from a stationary arrangement to one with moving bodies.

Elastic and inelastic collisions

In all collisions, momentum is conserved. However, there is generally a loss of kinetic energy in everyday collisions. The kinetic energy (KE) is transferred to different forms, such as heat, sound, or potential.

When there is a change in the KE, the collision is said to be *inelastic*.

In an *elastic* collision, KE is conserved (as well as momentum). No energy is lost or changed into other forms. Molecular collisions in gases are viewed as elastic collisions. Colliding snooker balls may be viewed as nearly elastic collisions, but obviously the collision is audible, so some KE has been transferred to sound.

Summary

- Momentum is defined as the product of the mass and velocity of a body.
- The principle of the conservation of momentum applies to all interactions provided no external resultant force acts.
- For a two-body collision: $m_1u_1 + m_2u_2 = m_1v_1 + m_2v_2$.

- Impulse is defined as the product of force and the time for which it acts.
- Impulse is equal to the change in momentum.
- Momentum is conserved in both elastic and inelastic collisions.

■ Work done, potential and kinetic energy

Work done

Work is done on a body when a force acting on the body makes it move. In doing so, energy is transferred to the body and takes a different form.

Doing work is a basic concept. Imagine stacking a shelf. It is clear to us that we are doing work, and the amount of work needed for the task depends on:

- the height of the shelf
- the weight of the object being lifted and hence the force needed to lift it

This forms the basis of the equation for work done:

work done = force × distance moved in the direction of the force

The unit of work is the joule (J).

Worked example

Figure 28 shows a force of 60 N acting on a paving slab resting on a horizontal surface. Find, by calculation, the work done in moving the slab 6.0 m along the surface.

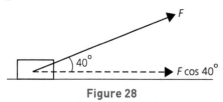

Figure 28

Answer

Find the component of the force acting in the direction of the movement, and multiply by the distance moved:

work done = $(F\cos\theta) \times d = (60\cos 40°) \times 6.0 = 276$ J

Energy transfer between gravitational potential energy and kinetic energy

In order to do work, or to perform a task, we require an energy source.

Consider our simple shelf-stacking example. Some chemical energy (derived from food) from the person stacking is converted into gravitational potential energy:

change in gravitational potential energy, $\Delta GPE = mg\Delta h$

If the object then fell from the shelf back to the floor, the gravitational potential energy would be converted or transferred, progressively, into the energy of movement, known as **kinetic energy**:

change in kinetic energy, $\Delta KE = \frac{1}{2}mv^2 - \frac{1}{2}mu^2$

Here u is the initial velocity, in this case zero, and v is the final velocity, in this case the velocity with which it hits the ground:

kinetic energy of an object (moving with velocity v), $KE = \frac{1}{2}mv^2$

The unit of any form of energy will be the same as that of work done — the joule, J.

Conservation of energy

This simple example illustrates a fundamental principle of physics known as the principle of the conservation of energy.

> One **joule** (1 J) of work is done when a force of one newton (1 N) moves its point of application by one metre (1 m), in the direction of the force.

> **Knowledge check 25**
>
> How much useful work is done by a horse in pulling a canal barge 120 m if the towrope makes an angle of 30° with the canal and the tension in the rope is 2200 N?

> **Energy** is the capacity to perform a task, while **work** is the 'doing' or carrying out of the task. The energy will change from one form into another.

> **Gravitational potential energy** is the energy possessed by an object due to its raised position above the Earth's surface.

> The **principle of the conservation of energy** states that energy can neither be created nor destroyed, but can be changed in form.

Worked example 1

A slate of mass 2.0 kg falls from a roof 7.0 m above the ground. Assuming it falls from rest, calculate its kinetic energy as it hits the ground and its speed.

Answer

$$\Delta KE = \Delta GPE$$

$$\Delta KE = mg\Delta h = 2.0 \times 9.81 \times 7.0 = 137.3\,J$$

$$\Delta KE = \tfrac{1}{2}mv^2 - \tfrac{1}{2}mu^2$$

In this example, $u = 0$, so

$$\Delta KE = \tfrac{1}{2}mv^2$$

$$137.3 = \tfrac{1}{2} \times 2.0 \times v^2$$

Therefore:

$$v^2 = 137.3$$

$$v = 11.7\,m\,s^{-1}$$

Worked example 2

The car at the top of one slope on a roller coaster is travelling at $4.0\,m\,s^{-1}$, as shown in Figure 29. It continues down the slope and up to the top of the next one. Ignoring losses due to friction, calculate the speed of the car at the top of the second slope.

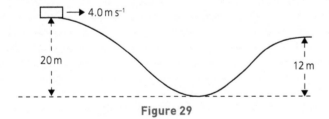

Figure 29

Answer

At the top of the first slope, E_{total} = GPE and KE.

At the bottom of the first slope, the GPE has been converted to an increased KE.

At the top of the second slope, some of this KE has been converted back to GPE.

Overall, there has been a net change of GPE to KE, because the second slope is lower:

$$\Delta GPE = \Delta KE$$

$$mg\Delta h = \tfrac{1}{2}mv^2 - \tfrac{1}{2}mu^2$$

The *m* cancels throughout:

$$g\Delta h = \tfrac{1}{2}v^2 - \tfrac{1}{2}u^2$$

$$9.81 \times (20 - 12) = \tfrac{1}{2}(v^2 - 16)$$

$$v^2 = 173.0$$

$$v = 13\,\mathrm{m\,s^{-1}}$$

Power

When work is done or a task completed, we are often concerned with how quickly it has been carried out. The rate at which the work is done defines **power**:

$$\text{power} = \frac{\text{work done}}{\text{time taken}}$$

This also means that the more power available, the quicker a task can be carried out.

The SI unit of power is the watt (W). Like energy, it is a scalar quantity. This means, for example, that a 5 W motor will transfer energy at the rate of 5 J every second.

There is another useful form of the power equation, found by substituting in the equation for work done:

$$\text{power} = \frac{F \times d}{t}$$

But:

$$\frac{d}{t} = v$$

Therefore:

$$\text{power} = \text{force} \times \text{velocity}$$

$$P = Fv$$

Worked example

An electric motor is used to pull a boat up a ramp. The tension in the connecting cable is 600 N. The boat moves up the ramp at a steady speed of 0.6 m s⁻¹. Find the power of the motor used.

Answer

$$P = Fv$$

$$P = 600 \times 0.6 = 360\,\mathrm{W}$$

Efficiency

When using devices that are doing work, it is important to know how much of the energy supplied is actually being converted into the required form for the job in hand, rather than wasted in the form of, for example, heat or sound.

The **efficiency** gives a quantitative measure of how much energy is changed into the correct form.

> **Knowledge check 26**
>
> With what speed will a can of beans hit the floor when it falls from a shelf 2.0 m above floor level?

> **Power** is defined as the rate at which work is done or energy is transferred from one form to another.

> **Exam tip**
>
> The letter W can stand for weight, work done or the unit watts. Symbols for variable quantities, such as weight or work done, are printed in *italics*, in this case *W*, while units are in normal text. So W (without italics) stands for watts. This difference is unseen when we record data, so be careful.

> **Efficiency** is the ratio of useful energy output to total energy input.

If we consider the energy output and input in one second, we have, in effect, a power ratio:

$$\text{efficiency} = \frac{\text{useful energy output}}{\text{total energy input}}$$

or

$$\text{efficiency} = \frac{\text{useful power output}}{\text{total power input}}$$

To obtain the percentage efficiency, we simply multiply the fractional efficiency by 100.

For a large electric motor, a typical efficiency would be 90%; whereas for a car engine, it would be 30%. Any device that runs at a higher temperature than its surroundings will lose considerable heat energy to the surrounding medium and consequently have a lower efficiency.

Worked example

A hoist raises a mass of 480 kg through 9.0 m by means of an electric motor. The mass is raised at a constant speed of $1.2\,\text{m s}^{-1}$.

a What is the gain in gravitational potential energy of the mass?

b How much mechanical power is the motor producing during the lift?

c If the motor is 90% efficient, how much electrical energy is used in the lift?

Answer

a The equation for change in gravitational potential energy
$\Delta\text{GPE} = mg\Delta h$ becomes:

$$\Delta\text{GPE} = 480 \times 9.81 \times 9.0 = 42\,379\,\text{J}$$

b Let P = the rate at which energy is transferred $= \dfrac{\Delta\text{GPE}}{t}$. Then:

$$\text{time} = \frac{\text{distance}}{\text{speed}} = \frac{9.0}{1.2} = 7.5\,\text{s}$$

$$P = \frac{42\,379}{7.5} = 5651\,\text{W}$$

$$\text{percentage efficiency} = \frac{\text{useful power out}}{\text{total power in}} \times 100$$

$$90 = \frac{5651}{\text{electric power supplied}} \times 100$$

$$P_{\text{electrical}} = \frac{5651}{90} \times 100 = 6278\,\text{W}$$

c From the main text above:

$$P = \text{rate at which energy is transferred}$$

Therefore:

$$E_{\text{electrical}} = P_{\text{electrical}} \times \text{time} = 6278 \times 7.5 = 47\,088\,\text{J}$$

(Note that 90% of the final answer equals the gain in gravitational potential energy.)

Exam tip

We could have used $P = Fv$, knowing the force to be equal in magnitude to the weight:
$P = (480 \times 9.81) \times 1.2 = 5651\,\text{W}$.

Knowledge check 27

A lift and its occupants weigh 12 000 N. What is the minimum power output of the motor needed to raise the lift at a steady speed of $2.0\,\text{m s}^{-1}$?

Knowledge check 28

A 5 W electric motor raises a weight of 15 N through a height of 2.0 m in 8.0 s. What is the efficiency of the motor?

Summary

- Work done is defined as the product of the force and the distance moved in the direction of the force.
- When work is done, energy is transferred from one form into another.
- The change in the gravitational potential energy of a body is given by the equation: $\Delta GPE = mg\Delta h$
- The change in the kinetic energy of a body is given by the equation:

  $\Delta KE = \frac{1}{2}mv^2 - \frac{1}{2}mu^2$

- The principle of the conservation of energy states that energy cannot be created nor destroyed, but can be changed from one form into another.

- Unlike momentum, kinetic energy is not conserved in all collisions. Kinetic energy is conserved in elastic collisions but not in inelastic collisions.
- Power is defined as the rate at which work is done or energy is transferred. It is given by the equations:

  $P = \dfrac{\Delta E}{t}$

  $P = Fv$

- The efficiency of a machine is a measure of how much of the energy supplied is transferred into the desired form:

  $\text{efficiency} = \dfrac{\text{useful energy out}}{\text{total energy in}}$

  It can also be expressed as a power ratio.

■ Electric current, charge, potential difference and electromotive force

Electrical energy is classed as high-grade energy because it is easily transferred into other useful forms. It was not until the creation of the National Grid in the 1930s that electricity became available to the majority of the population in England. Modern society is now totally reliant on electrical energy, and therefore an understanding of the fundamental concepts is useful.

Current

Moving electrical energy from one place to another, where it can be converted to mostly heat and light, relies on **electric current**, which flows through conducting wires.

Electric current can be expressed as an equation:

$I = \dfrac{\Delta Q}{\Delta t}$

where I is the current and ΔQ is the amount of charge passing in time Δt.

The unit of current is the ampere (A). The unit of charge is the coulomb (C). Therefore:

one ampere (1 A) = one coulomb per second (1 C s^{-1})

In metal conductors, the flowing charged particles are electrons, which are negatively charged.

An **electric current** is defined as the rate at which charged particles pass a point in a circuit.

Exam tip

Conventional current is denoted in circuit diagrams by an arrow in the direction from the positive terminal towards the negative terminal of the supply. Electrons move in the opposite direction.

The charge of a single electron is $e = 1.6 \times 10^{-19}\,\text{C}$. So the equation above becomes:

$$I = \frac{\Delta Q}{\Delta t} = \frac{Ne}{\Delta t}$$

where N is the number of electrons. So when a current of $1\,\text{A}$ is flowing in a wire, $1\,\text{C}$ of charge is passing each point per second, which means 6.25×10^{18} electrons pass each point in a second.

Potential difference

To make current flow between two points, a potential difference (p.d.) must exist between the two points. The defining equation is:

$$V = \frac{W}{Q}$$

where V is the potential difference, W is the work done or energy converted in the circuit and Q is the charge in coulombs.

Potential difference is measured in volts and, as can be seen from the equation above:

one volt $(1\,\text{V})$ = one joule per coulomb $(1\,\text{J}\,\text{C}^{-1})$

Electrical power and energy

The defining equation for potential difference can be adapted to give a relationship that links electrical power, current and the potential difference.

Rearranging the equations

$$I = \frac{Q}{t} \qquad \text{and} \qquad P = \frac{W}{t}$$

to give

$$Q = It \qquad \text{and} \qquad W = Pt$$

and substituting in

$$V = \frac{W}{Q} = \frac{Pt}{It} = \frac{P}{I}$$

we get

electrical power, $P = VI$

The unit for power, whether mechanical or electrical, is the watt (W).

Power, as stated, is the rate at which work is done or energy transferred:

$$P = \frac{W}{t}$$

Therefore, the energy transferred can also be expressed by the equation:

energy transferred, $W = Pt = VIt$

The unit for all forms of energy is the joule (J).

Knowledge check 29

The current in a certain wire is 1.5 A. Calculate
a the charge
b the number of electrons passing a point in the wire in 1 minute.

A potential difference is defined as the electrical energy converted per coulomb of charge passing between two points.

The unit of p.d. is the volt (V) and equals the p.d. between two points in a circuit in which one joule (1 J) of electrical energy is transformed when one coulomb (1 C) passes between the two points.

Knowledge check 30

A 3 kW immersion heater is designed for use on the 240 V mains.
a What is the current taken from the mains when it is operating?
b What energy, in MJ, does it use in 15 minutes?

Electromotive force

Batteries and generators supply energy to the charge as it passes through the source. The battery or generator is said to provide electromotive force, e.m.f.

The unit of e.m.f. is the volt (V). A car battery with an e.m.f. of 12 V supplies 12 J of energy per coulomb of charge passing through it.

Although e.m.f. and p.d. have the same unit, they are notably different. While the e.m.f. relates to a source and the energy it supplies, p.d. refers to electrical energy being converted in the circuit.

There is more about e.m.f on pp. 59–61.

Conservation laws applied to electrical circuits (Kirchhoff's laws)

Conservation of charge

Charge cannot physically leave a circuit, so the charge entering a junction, *per second*, is equal to the charge leaving the junction, *per second*. So the current into a junction is equal to the current out of the junction.

We can use the vector property of current here. Current in can be treated as positive and current out as negative, and so we can write:

$$I_1 + I_2 + I_3 = 0$$

See, for example, Figure 30.

Apply $I_1 + I_2 + I_3 = 0$ to Figure 30 to give:

$$2.5 + (-1.5) + I_3 = 0$$

Therefore $I_3 = -1.0\,A$, 1 A out of the junction.

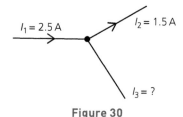

Figure 30

Conservation of energy

Energy conservation will apply to electric circuits as it does everywhere else.

Charge will move from an energy source and follow a circuit loop, returning back to the energy source to be re-energised. As it passes round, it will lose electrical energy to the components in the loop.

Imagine the journey of 1 C of charge round a circuit loop. The total electrical energy converted at the components in the loop, which is equal to the sum of the p.d.s round the loop (as $V = W/Q$), is equal to the total energy supplied to it, which is equal to the e.m.f. So:

$$\text{energy in} = \text{energy delivered or transferred}$$

$$\text{energy in per coulomb} = \text{energy delivered or transferred per coulomb}$$

$$\frac{E_{in}}{Q} = \frac{W}{Q}$$

e.m.f. $= \Sigma$ p.d.s

The e.m.f. *E* of a source is the energy converted into electrical energy when unit charge passes through it.

Exam tip

The e.m.f. and the terminal p.d. will have the same numerical value when the current drawn from the cell is zero.

Summary

- Electric current is the rate of flow of charge:
$$I = \frac{\Delta Q}{\Delta t}$$
- Potential difference measures the electrical energy transferred per coulomb of charge passing between two points:
$$V = \frac{W}{Q}$$
- Electrical power:
$$P = VI = I^2R = \frac{V^2}{R}$$
- Electrical energy transferred = $\frac{V}{t}$.

- The e.m.f., E, of a source is the energy converted into electrical energy when unit charge passes through it.
- At a junction in a circuit, the sum of the currents entering equals the sum of the currents leaving. This is a consequence of the law of the conservation of charge:
$$\Sigma\, I_{in} = \Sigma\, I_{out}$$
- In a circuit loop, the e.m.f. will equal the sum of the p.d.s.

■ Resistance and resistivity

Electrical resistance

Charges experience opposition to their movement as they move through conductors. This is called **resistance**; it gives a numerical value to the opposition and is defined in terms of the potential difference needed and the current it produces.

Resistance may be measured between two points in a circuit, but more commonly it is the resistance of a component in a circuit that is required:

$$\text{resistance, } R = \frac{V}{I}$$

The unit of resistance is the ohm (Ω), with $1\,\Omega$ being equivalent to $1\,VA^{-1}$.

Series and parallel circuits

To show how components are electrically connected to the electrical energy source, we draw circuit diagrams, using standard symbols. Components can be connected either in **series** or in **parallel**.

In the **series** arrangement, one component is followed by the next in a single path (Figure 31).

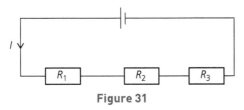

Figure 31

As a consequence, the current is the same at all points in a series circuit, and the potential differences across each component add up to the potential difference of the supply.

A numerical value of the **resistance** is defined as the ratio of the potential difference between two points to the current passing through those points.

Knowledge check 31

The current flowing in a light bulb is 3.0A when the p.d. is 12V. What is the resistance of the bulb?

Knowledge check 32

For a wire that carries 0.4A and has a resistance of 2Ω, determine the total energy dissipated in 5 minutes.

Exam tip

Make sure you know the symbols for battery, cell, variable power supply, switch, resistor, bulb, rheostat, thermistor, ammeter, voltmeter.

In a series circuit, the total resistance is equal to the sum of the individual resistances:

$$R_{total} = R_1 + R_2 + R_3$$

In a **parallel** arrangement, one component sits parallel to the next in a different loop; it is a multi-path circuit (Figure 32).

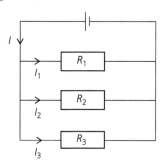

Figure 32

The current will divide at a junction, but, as no charge leaves the circuit, the sum of the currents in will equal the sum of the currents out of a junction:

$$I = I_1 + I_2 + I_3$$

The potential difference across each branch of a parallel arrangement is equal.

For resistors connected in parallel, the total resistance is given by the equation:

$$\frac{1}{R_{total}} = \frac{1}{R_1} + \frac{1}{R_2} + \frac{1}{R_3}$$

Most working circuits will be a combination of series and parallel sections.

Worked example 1

In the circuit in Figure 33, B is a 13 V battery of negligible internal resistance. The potential difference between X and Y is 5 V.

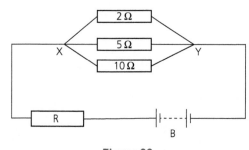

Figure 33

a Determine the effective resistance of the set of three parallel resistors between X and Y.
b What is the potential difference across the $2\,\Omega$ resistor?
c Determine the current flowing through the $10\,\Omega$ resistor.
d Determine the current flowing through the resistor R.
e Determine the magnitude of resistor R.

Knowledge check 33

What is the total resistance of three $12\,\Omega$ resistors placed **a** in series, **b** in parallel?

Answer

a For the three resistors connected in parallel:

$$\frac{1}{R_{effective}} = \frac{1}{R_1} + \frac{1}{R_2} + \frac{1}{R_3}$$

$$\frac{1}{R_{effective}} = \frac{1}{2} + \frac{1}{5} + \frac{1}{10}$$

$$\frac{1}{R_{effective}} = \frac{5}{10} + \frac{2}{10} + \frac{1}{10} = \frac{8}{10}$$

Therefore:

$$R_{effective} = \frac{10}{8} = 1.25\,\Omega$$

b The p.d. across each branch in a parallel section is the same:

$$V_{2\Omega} = V_{XY} = 5\,V$$

c $V_{10\Omega} = V_{XY} = 5\,V$

$$R = \frac{V}{I} \quad \text{or} \quad I = \frac{V}{R}$$

$$I_{10\Omega} = \frac{5}{10} = 0.5\,A$$

d Using the answers to parts a and b, we determine the current entering XY:

$$I_{XY} = \frac{V_{XY}}{R_{effective}} = \frac{5}{1.25} = 4.0\,A$$

This is also the value of the current through resistor R.

e $V_{supply} = V_R + V_{XY}$

$$V_R = 13 - 5 = 8\,V$$

$$R = \frac{V_R}{I_R} = \frac{8}{4} = 2\,\Omega$$

Worked example 2

Two identical bulbs are connected as shown in the circuit in Figure 34.

a If ammeter A1 reads 0.6 A, what are the readings on A2 and A3?

b If bulb X is now removed, what will be the readings on each of the three ammeters?

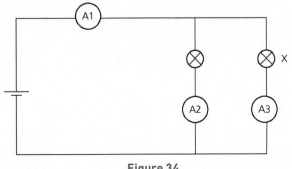

Figure 34

Answer

a Charge cannot be lost from a circuit. Therefore, there is conservation of charge and current at a junction. Let I_1, I_2 and I_3 be the currents measured by ammeters A1, A2 and A3. Then:

$$I_1 = I_2 + I_3 = 0.6\,\text{A}$$

As the bulbs have equal resistance, the current through each parallel branch will be equal:

$$I_2 = I_3 = 0.3\,\text{A}$$

The readings on A2 and A3 will both be 0.3 A.

b With bulb X removed, equivalent to a break in the circuit, no current can take this path, $I_3 = 0\,\text{A}$, and ammeter A3 will read 0 A.

However, the conditions in the other path have not changed from part a. It is the same bulb with the same resistance and the p.d. across the bulb is unchanged. Consequently, $I_2 = 0.3\,\text{A}$, and ammeter A2 will read the same as before, 0.3 A. Conservation of charge and thus current still applies:

$$I_1 = I_2 + I_3$$

$$I_1 = 0.3 + 0.0 = 0.3\,\text{A}$$

Ammeter A1 will read 0.3 A.

Exam tip

This question is often answered incorrectly. It is worth setting up the circuit to confirm the outcome. Common sense should also tell us that the current drawn in lighting one bulb should be half that for two.

Resistivity

The resistance of a wire at constant temperature depends on its dimensions, but also the material from which it is made. So there are three factors that determine the resistance of a material:

- length
- cross-sectional area
- type of material

Consider a wire:

- The longer the wire, the further the electrons must go and the greater the number of interactions with positive ions, resulting in a resistance increase. Experiment confirms that the resistance of a wire is proportional to its length:

$$R \propto L$$

- The thicker the wire, the greater the number of free electrons moving through the same distribution of positive ions, in effect a reduction in resistance. Experiment confirms that the resistance of a wire is inversely proportional to its cross-sectional area:

$$R \propto \frac{1}{A}$$

- The degree of interaction will differ depending on the wire type. Copper is a good conductor, while nichrome is a poor conductor.

Combining the first two experimentally confirmed relationships gives an overall equation for resistance, R:

$$R \propto \frac{L}{A} \qquad \text{or} \qquad R = \text{constant} \times \frac{L}{A}$$

We have yet to include the third factor — a value to reflect the wire type. This value is a property of the material from which the wire is made, and is a constant called its resistivity.

The resistivity ρ is a constant for a particular material, at constant temperature. It is a measure of the opposition to charge flow due to the nature or type of the material. So we have

$$R = \text{constant} \times \frac{L}{A} = \frac{\rho L}{A}$$

and therefore

$$\rho = RA/L$$

So the value of ρ is equal to R when $L = 1\,$m and $A = 1\,$m^2.

The units of ρ are ohm metres (Ωm).

The value of the resistivity of a material varies greatly, from an order of magnitude of $10^{-8}\,\Omega$m for a conductor to $10^{13}\,\Omega$m for an insulator.

Experimental determination of resistivity

Finding the resistivity of a wire involves the use of the two-meter method — ammeter and voltmeter — to determine resistance, and measurement of the dimensions of the wire. In keeping with good experimental practice, a series of results should be obtained. For example, for a wire of fixed thickness, you might find the resistance of several lengths, from 30 cm to 100 cm.

The test circuit used for a series of current–voltage values, with the wire placed in the component position, is shown in Figure 35.

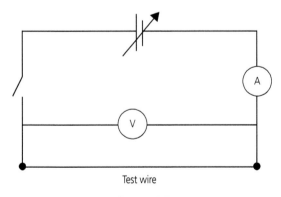

Test wire

Figure 35

Exam tip

Resistivity is to resistance what the Young modulus is to the force constant. The resistance and the force constant will have different values for different wires of the same material, whereas the resistivity and Young modulus are constants for the material, no matter what its dimensions.

Knowledge check 34

Calculate the resistivity of the material used for a heating element of length 0.8 m with cross-sectional area 0.06 mm^2 and resistance 15 Ω.

Exam tip

Stretch the wire first to remove any kinks, and then attach it along the length of a metre rule. Use a low voltage setting on the power pack to ensure that the current is low and the temperature variation is minimal.

- Using a micrometer screw gauge, measure the diameter of the wire at three different positions and calculate the average diameter $\langle d \rangle$. To calculate the cross-sectional area, use the equation:

$$A = \frac{\pi \langle d \rangle^2}{4}$$

- To calculate the resistance of each length of wire, use:

$$R = \frac{V}{I}$$

- Plot a graph of resistance R against the length L.
- Find the gradient of the best-fit straight line for the points plotted (Figure 36).

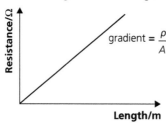

Figure 36

Using the equation:

$$R = \frac{\rho L}{A}$$

and rearranging gives:

$$\frac{\rho}{A} = \frac{R}{L} = \text{gradient}$$

Therefore:

$$\rho = \text{gradient} \times A$$

Worked example

The heater for the rear window of a car consists of two strips of good electrical conductor joined by six thin strips of resistive metal (Figure 37). Each thin strip is 0.80 m long, with cross-sectional area 0.12 mm², and is made from material of resistivity $4.5 \times 10^{-6}\,\Omega\,\text{m}$ at room temperature.

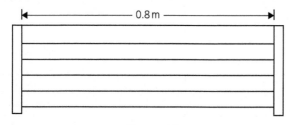

Figure 37

a Calculate the resistance of one of the strips at room temperature.
b Calculate the resistance of the complete heater at room temperature.
c Taking the terminal p.d. of the car battery to be 12 V, what power is delivered by the battery to the heater?
d Why are the strips connected as shown?

Exam tip

Be careful with your units. It is best to work in metres throughout. When converting mm² to m², divide by 10^6.

Answer

a The resistance R of one strip is:

$$R = \frac{\rho L}{A}$$

$$R = \frac{(4.5 \times 10^{-6}) \times 0.8}{0.12 \times 10^{-6}}$$

$$R = 30\,\Omega$$

b There are six *equal* resistors in parallel, each strip with resistance R:

$$\frac{1}{R_{total}} = \frac{1}{R_1} + \frac{1}{R_2} + \frac{1}{R_3} + \frac{1}{R_4} + \frac{1}{R_5} + \frac{1}{R_6}$$

$$\frac{1}{R_{total}} = \frac{6}{R}$$

$$\frac{1}{R_{total}} = \frac{6}{30}$$

$$R_{total} = 5\,\Omega$$

c Power is given by:

$$P = VI = I^2R = \frac{V^2}{R}$$

$$P = \frac{12^2}{5} = 28.8\,\text{W}$$

d They are connected in parallel so that, if one strip breaks, the others continue to operate.

Current–voltage characteristics

A variety of components are used in electrical circuits, for example resistors, lamps, motors, thermistors, diodes and transistors. The behaviour or characteristics of these are best displayed in a current–voltage graph for each component.

The circuit in Figure 38 is used to test a component. By varying the supply p.d., a range of current and p.d. values can be recorded for the component using the ammeter and voltmeter.

Note that the supply can be reversed to get the negative p.d. values.

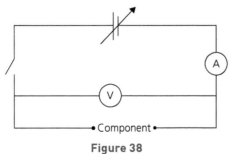

Figure 38

The results are displayed for each component in a current–voltage graph.

Figures 39–42 show the current–voltage characteristics for different components in a series of graphs.

Figure 39 could be a wire resistor. The current and potential difference are directly proportional to each other when the current flows in either direction.

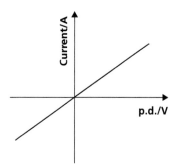

Figure 39 Metallic conductor at constant temperature

Figure 40 could be the metal filament in a bulb. The current and potential difference are not proportional because the current heats the filament. An increase in the temperature of the wire increases the resistance of the filament.

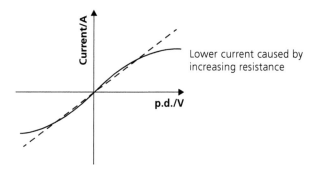

Figure 40 Metallic conductor *not* at a constant temperature.

In Figure 41 there is almost no current when the p.d. is applied in one direction, called 'reverse-biased'. When the p.d. is applied in the other, 'forward-biased', direction, the current will increase rapidly, but only after a small 'trigger voltage' has been passed.

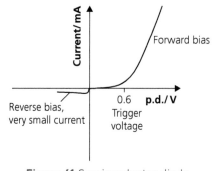

Figure 41 Semiconductor diode

The resistance of a thermistor changes rapidly with temperature. For a negative-temperature-coefficient (NTC) thermistor semiconductor, the increase in temperature, and with it thermal energy, causes a significant increase in the number of conducting electrons and so an increased current value (Figure 42).

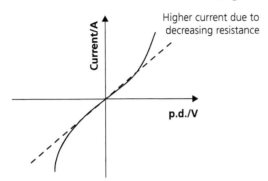

Figure 42 A negative-temperature-coefficient thermistor

Ohm's law

The current–voltage characteristic shown in Figure 39 is referred to as ohmic behaviour, and a conductor that displays this proportionality is an ohmic conductor. Ohmic conductors are conductors that obey Ohm's law.

To identify a true non-ohmic conductor, we need to see a current–voltage characteristic whose non-linear appearance is not a consequence of temperature change, for example a semiconductor diode.

Resistance and temperature

Electrical conductors have almost free, delocalised electrons, which flow under the influence of a potential difference. This is in contrast to insulators, where the electrons are tightly held to an individual atom and therefore there are no charge carriers and no current is possible.

A conductor also has vibrating positive ions. As the electrons flow, they interact with these ions; it is this impedance to charge flow that constitutes resistance.

As the temperature of a metal increases, the positive ions vibrate with a greater amplitude. This results in greater opposition to electron flow and an increase in the resistance.

With a thermistor — a semiconductor — there is a second, more dominant, effect. The increase in temperature also results in the release of more conduction electrons, and therefore there is an overall decrease in the resistance.

This variation of resistance with temperature change can be confirmed experimentally and results displayed in graphical form (Figures 43 and 44).

Ohm's law states that the current through a metallic conductor is directly proportional to the potential difference across it, provided the temperature remains constant.

Exam tip

Metals are ohmic conductors, yet the metal in a filament bulb — tungsten — does not display the proportionality. This is because the condition of having a constant temperature does not apply. The filament temperature will rise from room temperature to about 1500°C.

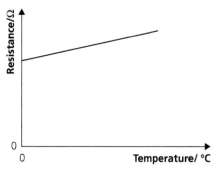

Figure 43 Resistance change of a conductor against temperature

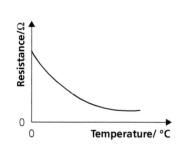

Figure 44 Resistance change of a semiconductor against temperature

Exam tip

Resistance should always be calculated from discrete values of *V* and *I*. It will only equal the inverse of the gradient of the *I* versus *V* graph if this is a straight line.

Experiment to show the variation with temperature of the resistance of an NTC thermistor

Two different methods are described, the main difference being that, in one, the data are obtained as the thermistor is cooling, while, in the other, data are obtained as it is being heated.

The merits of one method compared with the other can be discussed. Having thermal equilibrium between the thermistor and the liquid in which it is placed is important.

Method A

1 Connect the thermistor to a multimeter set to record resistance, an ohmmeter.
2 Record the value of the resistance of the thermistor and the temperature of the room in which it is sitting.
3 Place the thermistor into a beaker.
4 Half fill the beaker with boiling water, ensuring that the thermistor is totally covered by water.
5 Place the thermometer into the water. The arrangement is shown in Figure 45.
6 After stirring the water, record the temperature of the water and the resistance of the thermistor.
7 Repeat step 6 every 10° drop in temperature (the process may be speeded up by adding some ice).
8 Plot a graph of thermistor resistance against temperature.

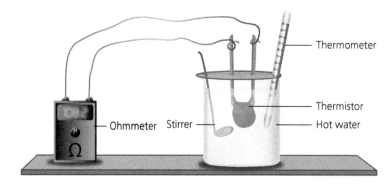

Figure 45

Method B

1 Connect the thermistor to a multimeter set to record resistance, an ohmmeter.
2 Place the thermistor into a test tube of glycerol, which is in turn placed into a beaker of water.
3 Suspend a thermometer in the glycerol and measure the temperature.
4 Record the resistance of the thermistor before heating commences.
5 Gradually heat the water using a Bunsen burner. The arrangement is shown in Figure 46.
6 After a rise of approximately 10°, remove the heat, wait until the temperature stops rising and then record both the resistance and the temperature.
7 Recommence the heating and repeat step 6 until the water boils.
8 Plot a graph of thermistor resistance against temperature.

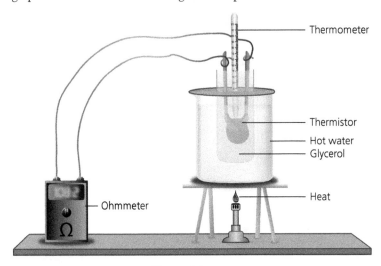

Figure 46

The resulting graphs should look like that shown in Figure 44.

Uses of NTC thermistors

NTC thermistors are used as resistance thermometers in low-temperature measurements of the order of 10 K. A calibration experiment, like that described above, is carried out so that any resistance can be equated to a temperature. This type of temperature measuring device has the advantages of allowing remote recording in very low-temperature environments and a rapid response to temperature change.

NTC thermistors can be used as anti-surge current-limiting devices in power supply circuits. They present a higher resistance initially, which prevents large currents from flowing at turn-on, and then heat up and become much lower resistance to allow higher current flow during normal operation.

NTC thermistors are regularly used to monitor things like coolant temperature and/or oil temperature inside a car engine and provide data to the control unit and, indirectly, to the dashboard.

NTC thermistors can also be used to monitor the temperature of an incubator.

NTC thermistors are used in fire alarm circuits. When the temperature goes past a set temperature, the current through the thermistor is at a value to trigger a relay circuit, which rings an alarm.

Superconductivity

The resistance that we have described in conductors due to the interaction between the electrons and the vibrating ions is seen to reduce as the temperature of the conductor becomes lower. If this temperature reduction is continued to temperatures near absolute zero, then the vibration of the ions is almost zero and there should be little opposition to electron movement. (The cooling of the materials is achieved using liquid nitrogen or liquid helium.)

With further low-temperature investigation, it has been found that, for several metals, a condition of zero resistance is achieved, not gradually, but abruptly at a temperature now called the **critical temperature**, T_c. Materials that display this characteristic are called superconductors. Some ceramic materials have also been found to display the property of superconductivity — and at higher temperatures, for example, at about 90 K.

Below the critical temperature, the superconducting materials have no electrical resistance, and so they can carry large electric currents without losing energy as heat (Figure 47).

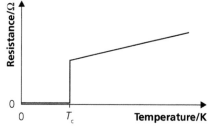

Figure 47 Resistance change of a superconductor against temperature

Superconductors that provide large currents with little energy loss are used in the powerful electromagnets required in magnetic resonance imaging (MRI) and accelerator technology.

Exam tip

The critical temperature is also referred to as the transition temperature.

Summary

- The resistance of any component:

$$R = \frac{\text{p.d. across the component}}{\text{current through it}} = \frac{V}{I}$$

- Resistance depends on the type and dimensions of the conductor.
- Conventional current direction is taken as from the positive terminal towards the negative terminal of the supply.
- The total resistance of any number of resistors in series is given by:

$$R_{\text{total}} = R_1 + R_2 + R_3 + \dots$$

- The total resistance of any number of resistors in parallel is given by:

$$\frac{1}{R_{\text{total}}} = \frac{1}{R_1} + \frac{1}{R_2} + \frac{1}{R_3} + \dots$$

- Resistivity ρ is a measure of a material's opposition to current and, unlike resistance, is independent of the dimensions of the sample:

$$\rho = \frac{RA}{L}$$

- The unit of resistivity is the ohm metre (Ω m).
- Ohm's law states that the current through a metallic conductor is directly proportional to the potential difference across it, provided the temperature remains constant.
- The resistance of a metallic conductor will increase as its temperature increases.
- The resistance of a negative-temperature-coefficient (NTC) thermistor will decrease as the temperature of the thermistor increases. Thermistors are used as temperature sensors.
- A superconductor has zero resistance below a critical temperature.
- Superconductors are used to deliver the large currents needed to make high-power electromagnets.

Internal resistance and electromotive force

All electrical power sources have some resistance of their own, called the internal resistance.

Each charge gains energy as it moves through the source, but some of the energy is used up in pushing the charge through the source. This energy is dissipated as heat, which is why a battery can become warm when supplying current.

We can treat the internal resistance, r, as a small resistor in series with the supply (Figure 48).

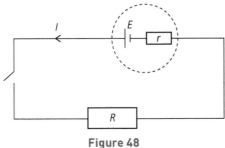

Figure 48

For a supply with e.m.f. E and internal resistance r, delivering current I to a circuit of resistance R:

$$E = IR + Ir$$

Here $IR = V_{terminal}$ is the potential difference available to the circuit. $Ir = V_{lost}$ is the potential difference used to push the charge through the cell.

The above equation can be written in different forms:

$$E = V_{terminal} + V_{lost}$$

$$E = V_{terminal} + Ir$$

$$E = IR + V_{lost}$$

The effects of the internal resistance are as follows:

- When $I = 0\,A$, apply $E = V_{terminal} + Ir$, so $E = V_{terminal}$.
- When I is high, apply $E = V_{terminal} + Ir$, so $V_{terminal}$ will be low.

$V_{terminal}$ is the terminal potential difference, the p.d. between the terminals of a battery or cell when current is being delivered.

Worked example

What is the terminal p.d. of a car battery of e.m.f. $12\,V$ and $r\ 0.04\,\Omega$ when it is delivering $100\,A$ to the starter motor?

Answer

Rearranging the equation in the text:

$$E = V_{terminal} + Ir$$

$$V_{terminal} = E - Ir = 12 - (100 \times 0.04) = 8\,V$$

(It is this drop in terminal p.d. that is responsible for the lights of a car dimming as the engine is turned on.)

Exam tip

When recharging a battery, the potential difference applied across its terminals must be greater than the e.m.f.

Experimental determination of the internal resistance

The circuit in Figure 49 can be used to measure the e.m.f. and the internal resistance of a cell.

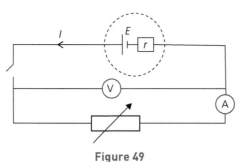

Figure 49

- Adjust the variable resistor to its maximum resistance value.
- Then close the switch and immediately record the values of the terminal p.d. and current from the voltmeter and ammeter respectively.
- Reopen the switch and decrease the resistance of the variable resistor.
- Close the switch again and record the new values on the voltmeter and ammeter.
- Repeat this process several times and then plot the terminal p.d. on the y-axis against current on the x-axis (Figure 50).

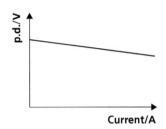

Figure 50

Compare the equation for the e.m.f. and the general equation for a straight-line graph :

$$V_{\text{terminal}} = -Ir + E \qquad \text{and} \qquad y = mx + c$$

We see that:

y-axis intercept, $c = E$, e.m.f.

gradient, $m = -r$, internal resistance

Summary

- For a supply with e.m.f. E and internal resistance r, $V = E - Ir$, where V is the terminal potential difference available to the circuit.
- The terminal p.d. will always be less than the e.m.f. when the supply is delivering current (due to internal resistance).
- Dry cells have a significant internal resistance, whereas a car battery must have a low internal resistance, as it needs to produce a large current for the starter motor.

■ Potential divider circuits

There are occasions when only a fraction of the p.d. provided by a source is required, so we need to tap part of the p.d. available. This can be achieved by using two resistors in series with the supply (Figure 51).

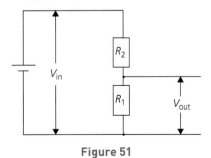

Figure 51

In this potential divider circuit with two resistors:

$$V_{in} = I(R_1 + R_2)$$

$$V_{out} = IR_1$$

where I is the current through the two resistors in series with the supply.

Combining the two equations gives:

$$\frac{V_{in}}{R_1 + R_2} = I = \frac{V_{out}}{R_1}$$

Rearranging:

$$\frac{V_{out}}{V_{in}} = \frac{R_1}{R_1 + R_2}$$

$$V_{out} = \frac{R_1 V_{in}}{R_1 + R_2}$$

> **Exam tip**
>
> When using the equation in this form, note the relative positions of R_1 and R_2 in the circuit diagram and in the equation. It is best to be able to generate the equation from first principles.

Variable potential

On occasions we may want a continuous range of p.d.s, for example in a school variable power supply or in a lighting dimmer circuit. This is easily obtained by replacing the two discrete resistors R_1 and R_2 by one slide variable resistor. Examples are illustrated in Figure 52.

> **Exam tip**
>
> This circuit has the advantage over simply placing a light in series with the variable resistor in that zero p.d. can be achieved with the potential divider but not with the simpler circuit.

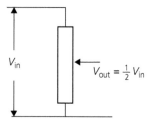

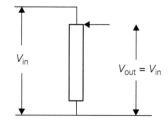

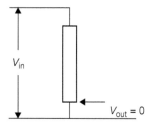

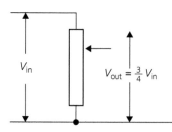

Figure 52

> **Exam tip**
>
> When a variable resistor is used to control current in a circuit, we refer to it as a rheostat. When a variable resistor is used to control voltage in a circuit, we refer to it as a potential divider.

Loading the potential divider

The component or load, for example a bulb, is placed across the V_{out} terminals (Figure 53).

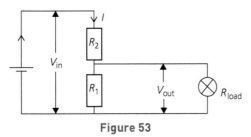

Figure 53

The V_{out} that will be predicted by use of the equation is theoretical.

Placing a load in parallel with R_1 will reduce the effective resistance of this part of the circuit and consequently the p.d.

Before it is applied, the effective resistance R_{eff} of R_1 and R_{load} should be calculated and used as a replacement for the R_1 value in the equation:

$$\frac{1}{R_{eff}} = \frac{1}{R_1} + \frac{1}{R_{load}}$$

$$V_{out} = \frac{R_{eff}V_{in}}{R_{eff} + R_2}$$

Worked example

a Calculate the potential difference between X and Y in Figure 54.

b Calculate the potential difference between X and Y when *another* $4\,\Omega$ resistor is connected between these points.

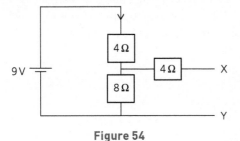

Figure 54

Answer

a The equation in the text gives

$$V_{out} = \frac{R_1 V_{in}}{R_1 + R_2}$$

$$V_{XY} = \frac{8 \times 9}{4 + 8} = 6\,V$$

b This situation is shown in Figure 55.

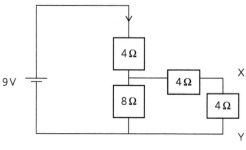

Figure 55

For the two $4\,\Omega$ resistors in series, total resistance $R_L = 4 + 4 = 8\,\Omega$.

For this $R_L = 8\,\Omega$ in parallel with R_1, of value $8\,\Omega$:

$$\frac{1}{R_{eff}} = \frac{1}{R_1} + \frac{1}{R_L}$$

$$\frac{1}{R_{eff}} = \frac{1}{8} + \frac{1}{8} = \frac{2}{8} = \frac{1}{4}$$

$$R_{eff} = 4\,\Omega$$

Thus:

$$V_{out} = \frac{R_{eff}V_{in}}{R_{eff} + R_2}$$

$$V_{XY} = \frac{4 \times 9}{4 + 4} = 4.5\,V$$

Sensors as part of the potential divider circuit

Thermistors and light-dependent resistors (LDRs) can be incorporated into potential dividers to cause a variation in the output voltage, which in turn can be used to trigger the operation of lighting or heating circuits. (The second resistor can be a variable resistor, so acting as a sensitivity control.)

Heater circuit

One of the two resistors in the potential divider is replaced by a thermistor. As the temperature surrounding the thermistor falls, its resistance increases and V_{out} increases to a value needed to switch on the heater (Figure 56).

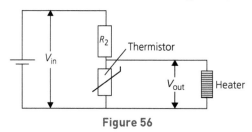

Figure 56

Exam tip

The LDR has a low resistance in bright light and a high resistance in darkness.

Lighting circuit

One of the two resistors in the potential divider is replaced by an LDR. As the light intensity surrounding the LDR falls, its resistance increases and V_{out} increases to a value needed to switch on the light (Figure 57).

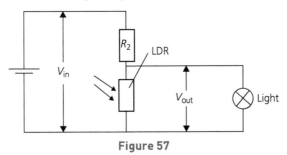

Figure 57

Summary

- Two or more resistors connected in series to a supply act as a potential divider.
- For a two-resistor potential divider:

$$V_{out} = \frac{R_1 V_{in}}{R_1 + R_2}$$

- The potential difference output of a potential divider circuit will drop once it is connected to a load.

- Variable potential dividers can be used in volume control and dimmer circuits, to allow variation from zero to a maximum.
- Variable potential dividers can be used in sensor circuits to activate lighting and heating.

Questions & Answers

The unit assessment

Unit AS 1 is a written examination consisting of a number of compulsory short-answer questions. Some of the questions will require extended responses and most will contain a mathematical calculation component.

The maximum mark for the paper is 100 and the duration of the examination is 1 hour 45 minutes. Unit AS 1 contributes 40% to the overall AS award and 16% of the overall A-level.

You should ensure that your writing is legible and that the meaning is clear and relevant. Select a form of writing appropriate to the purpose, in a logical sequence, and include scientific vocabulary where appropriate.

Command terms

Examiners use certain words to indicate the type of response expected. (The depth of response required is usually indicated by the number of marks allocated.) It is helpful to be familiar with these terms.

- **Define** — a formal statement. If an equation is given, each term used must be identified.
- **State** — a concise answer with little or no supporting argument.
- **Explain** — usually a definition together with some relevant comment on the significance of the term.
- **List** — a number of points or terms with no explanation.
- **Describe** — state the key points relating to a particular concept or experiment.
- **Calculate** — used when a numerical answer is required. Working out should be shown.
- **Measure** — indicates that a quantity can be obtained directly by use of a suitable piece of apparatus.
- **Determine** — indicates that a quantity cannot be obtained directly but rather by calculation, with given or measured values substituted in a suitable form of a known equation.
- **Show** — used when a calculation has to be performed to obtain an already given value. All work must be shown clearly and logically, with the final value shown to a greater significance than required to confirm that it has not been a fudge.
- **Sketch** — when related to graphs, you are expected to show the general shape or trend of the graph, including intercepts if appropriate. The detail on the axes — quantities, units, origin — is normally expected. Additional values will be specifically asked for if required.
- **Draw** — a carefully and fully labelled diagram, with apparatus shown, using standard symbols, to enable all necessary measurements to be taken. Use a ruler where appropriate and do not make it small. Note the difference between a circuit diagram and a drawing of the apparatus required.

Data and formulae sheet for Unit AS 1

A data and formulae sheet will be provided inside the AS 1 examination paper. The constants given may be required to allow you to complete some of the calculations. Note that all the values are given to three significant figures. The following values and formulae are provided:

- speed of light in a vacuum $\quad\quad\quad\quad\quad\quad c = 3.00 \times 10^8\,\mathrm{m\,s^{-1}}$
- elementary charge $\quad\quad\quad\quad\quad\quad\quad\quad\quad e = 1.60 \times 10^{-19}\,\mathrm{C}$
- Planck constant $\quad\quad\quad\quad\quad\quad\quad\quad\quad\quad h = 6.63 \times 10^{-34}\,\mathrm{J\,s}$
- mass of electron $\quad\quad\quad\quad\quad\quad\quad\quad\quad\quad m_e = 9.11 \times 10^{-31}\,\mathrm{kg}$
- mass of proton $\quad\quad\quad\quad\quad\quad\quad\quad\quad\quad m_p = 1.67 \times 10^{-27}\,\mathrm{kg}$
- acceleration of free fall on Earth's surface $\quad g = 9.81\,\mathrm{m\,s^{-2}}$
- electron volt $\quad\quad\quad\quad\quad\quad\quad\quad\quad\quad\quad 1\,\mathrm{eV} = 1.60 \times 10^{-19}\,\mathrm{J}$
- conservation of energy $\quad\quad\quad\quad\quad\quad \frac{1}{2}mv^2 - \frac{1}{2}mu^2 = Fs$ (for a constant force F)
- terminal potential difference $\quad\quad\quad\quad V = E - Ir$ (e.m.f. E; internal resistance r)
- potential divider $\quad\quad\quad\quad\quad\quad\quad\quad V_{\mathrm{out}} = \dfrac{R_1 V_{\mathrm{in}}}{R_1 + R_2}$

Revision tips

- The examination is designed to test all the content of the unit, and, as all the questions are compulsory, your revision must address every element.
- Break down and learn in detail the content stated in the specification and covered in the Content Guidance section of this guide:
  - experiment descriptions
  - full definitions
  - equations, in all forms
  - statements of laws and principles
- Sequence your notes and corrected example questions in the order of the specification.
- Use past paper questions and those provided in this guide to ensure that you are familiar with typical questions associated with each topic.

About this section

This section consists of two self-assessment tests, each of 100 marks in total, as in the real exam. Allow 1 hour 45 minutes if completing one of these tests in one sitting.

First, try the questions without reference to the answers. Then compare your responses with the given solutions and read the comments to give you an added insight into each problem and the probable pitfalls.

For question parts worth multiple marks, ticks (✓) are included in the answers to indicate where marks have been awarded.

Comments on some questions are preceded by the icon ⓔ. They offer tips on what you need to do to gain full marks. Some answers are followed by comments, indicated by the icon ⓔ, which highlight where credit is due or could be missed.

■ Self-assessment test 1

Question 1

(a) Show that the base units of potential energy and kinetic energy are the same.　　(4 marks)

Ⓔ Be clear of the difference between the SI unit for the quantity and the base units. The SI unit for any form of energy is the joule. Note that the SI unit of mass is the kilogram and not the gram.

(b) A mass m is attached to one end of a suspended spiral spring. This produces a force F in the spring. The mass is then pulled down and released, causing it to oscillate. Equation 1 represents the relationship for the periodic time T of mass–spring oscillation.

$$T = 2\pi\sqrt{\frac{my}{F}} \qquad \text{(Equation 1)}$$

Determine the base units of the term y.　　(2 marks)

Answer to Question 1

(a) potential energy = mgh ✓ where g is the acceleration due to gravity.

units of mass = kg

units of acceleration = $m\,s^{-2}$

units of height = m

base units of potential energy = (kg) × ($m\,s^{-2}$) × (m) = $kg\,m^2\,s^{-2}$ ✓

kinetic energy = $\frac{1}{2}mv^2$ ✓

units of mass = kg

units of velocity = $m\,s^{-1}$

base units of kinetic energy = (kg) × ($m\,s^{-1}$)^2 = (kg) × ($m\,s^{-1}$) × ($m\,s^{-1}$)

　　　　　　　　= $kg\,m^2\,s^{-2}$ ✓

Both forms of energy have the base units $kg\,m^2\,s^{-2}$.

Ⓔ Make sure you show all stages of the solution to convince the examiner that your answer is not a bluff.

(b) base units of periodic time T = s

base units of mass m = kg

base units of force F = $kg\,m\,s^{-2}$ ✓

Ⓔ It is often simpler to remove a square root by squaring both sides of the equation.

Squaring the equation:

$$T^2 = 4\pi^2 \frac{my}{F}$$

Rearrange the equation with y as the subject:

$$y = \frac{T^2 F}{4\pi^2 m}$$

Enter the base units of the quantities:

base units of $y = (s^2) \times (kg\,m\,s^{-2})/(kg) = m$ ✓

ⓔ In general, when a derived unit is involved in a question — for example, for power, watts (W) — be sure to be working in the SI base units for mass, length, time and temperature.

Question 2

(a) State the difference between a scalar and a vector quantity. (1 mark)

(b) Indicate which of the four physical quantities listed below are **scalars** by underlining them. (2 marks)

velocity   frequency   kinetic energy   density

(c) A ball is kicked towards a wall with a velocity of $3.5\,m\,s^{-1}$ and rebounds directly back at the same speed. What is the change in the velocity of the ball? (2 marks)

ⓔ When asked about a difference, always make reference to both quantities.

Answer to Question 2

(a) Vector quantities have a directional aspect; scalars do not. ✓

(b) frequency   kinetic energy   density ✓✓ ($\frac{1}{2}$ mark each, round up; penalty of 1 mark if velocity underlined)

(c) While speed has not changed, the direction has changed; the velocity after hitting the wall is $-3.5\,m\,s^{-1}$. ✓

change in velocity = velocity after rebound – velocity before

change in velocity = $(-3.5) - (3.5) = -7.0\,m\,s^{-1}$ ✓

ⓔ Note that the negative in the answer would not be a marking point on this occasion, as only the *change* was asked for. The answer should be quoted to the same number of significant figures as the data given.

Question 3

The diagram below shows a force of 11 N acting on a block resting on a horizontal surface.

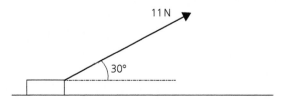

(a) Find the horizontal and vertical components of the force.   (2 marks)

(b) What is the resultant vertical force acting on the horizontal surface if the block has a mass of 4.0 kg?   (2 marks)

> **Answer to Question 3**
>
> **(a)** horizontal component of the force = 11 cos 30° = 9.5 N ✓
>
> vertical component of the force = 11 sin 30° = 5.5 N ✓
>
> **(b)** weight = mg = 4.0 × 9.81 = 39.2 N ✓
>
> resultant force = weight − vertical component of the force
>
> resultant force = 39.2 − 5.5 = 33.7 N or 34 N ✓

ⓔ Use the value shown on the calculator from earlier question parts and not rounded values when substituting into a later equation. Then carry out one final rounding to the appropriate significance at the end (in this case two significant figures).

Question 4

A car is stopped at traffic lights. The lights change to green and, just as the car starts to move, a bus, in the bus lane, passes it moving at a steady velocity of 15.0 m s⁻¹. The car accelerates at a steady rate, reaching a velocity of 22.5 m s⁻¹ in 30 s and then maintains this velocity.

(a) Taking it that the lights become green at time = 0, draw the velocity–time graphs for the car and the bus up to t = 60 s. Use the same axes for both.   (3 marks)

(b) How long does it take the car to reach the same velocity as the bus?   (1 mark)

(c) At that time, what is the distance between the car and the bus?   (2 marks)

(d) What is the acceleration of the car in the time interval 0–30 s?   (2 marks)

(e) How long after the lights change does the car pass the bus?   (3 marks)

Answer to Question 4

(a) See the graph below.

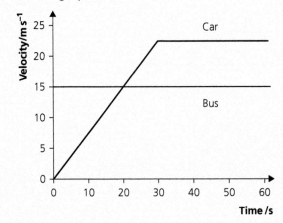

graph/scales ✓

bus — horizontal line ✓

car — diagonal and horizontal ✓

ℯ The only odd number that is acceptable to use in scaling is 5. Be sure to give the units and label in the correct format, 'Quantity/unit', to give a pure number. Graphs only deal with pure numbers.

(b) Intersection of the two plots at 20 s ✓

(c) Area between the graphs and the time axis up to 20 s gives the distance travelled. ✓

for the bus: $20 \times 15 = 300\,\text{m}$

for the car: $\frac{1}{2}(20 \times 15) = 150\,\text{m}$

150 m apart ✓

(d) Acceleration is the slope of the velocity–time plot. ✓

$$a = \frac{\Delta v}{\Delta t} = \frac{22.5}{30} = 0.75\,\text{m s}^{-2} \checkmark$$

(e) Find the distance travelled by both after 30 s.

for the bus: distance $= 30 \times 15 = 450\,\text{m}$

for the car: distance $= \frac{1}{2}(30 \times 22.5) = 337.5\,\text{m}$

112.5 m apart ✓

From now on, the car catches up on the bus by $(22.5 - 15) = 7.5\,\text{m}$ each second. ✓

Therefore, it will take a further $\dfrac{112.5}{7.5} = 15\,\text{s}$ to pass the bus.

This is 45.0 s after the lights changed. ✓

e Part (e) is an unstructured, multi-step question. It is important to show as much of each step as you can. If you know something about the latter stage only, show it. Error carried forwards can be applied within a multi-step answer.

Question 5

An aircraft at a height of 1.2×10^4 m is travelling horizontally at $360 \, km \, h^{-1}$. When the aircraft is vertically above a target, a bomb is released.

(a) How long before the bomb hits the ground? (3 marks)

(b) How far from the target does the bomb hit the ground? (4 marks)

Answer to Question 5

(a) The bomb has no initial vertical component of velocity. ✓

Use $s = ut + \frac{1}{2}at^2$: ✓

$$t^2 = \frac{2s}{a} = \frac{2 \times 1.2 \times 10^4}{9.81} = 2446$$

$t = 49.5 \, s$ ✓

(b) Convert $km \, h^{-1}$ to $m \, s^{-1}$:

$$\frac{360 \times 1000}{60 \times 60} = 100 \, m \, s^{-1} \checkmark$$

The bomb has a horizontal velocity equal to that of the aircraft. ✓

There is no horizontal acceleration. ✓

Use $s = ut + \frac{1}{2}at^2$; which reduces to $s = ut$:

$s = (100 \times 49.5) = 4950 \, m$ ✓

---

e Note that 50% of the marks in part (b) are for comments on the principles, without calculation. This again highlights that it is worth recording all you know about the answer to the question.

Question 6

(a) State Newton's second law of motion. (2 marks)

(b) When a car engine is providing a driving force of 700 N, it moves along a level road with an acceleration of $0.4 \, m \, s^{-2}$.
The driving force is then increased to 1200 N, and the new acceleration is $0.8 \, m \, s^{-2}$.
Assume the frictional force on the car remain constant.

    (i) Calculate the value of the frictional force. (4 marks)

    (ii) Determine the mass of the car. (1 mark)

ⓔ Always read the full question. The overall context and specific clues to the relevant equations will appear in the wording; in this case the forces involved — the driving force and the frictional force.

Answer to Question 6

(a) For a body of constant mass, its acceleration is directly proportional to the resultant force applied to it ✓ and in the direction of the resultant force. ✓

ⓔ Newton's second law can be expressed in terms of 'rate of change of momentum'. Minor word variations are acceptable, for example net force rather than resultant force.

(b) (i) Use the equations:

$$F_{resultant} = ma \quad \text{and} \quad F_{resultant} = F_{driving} - F_{friction}$$

Therefore:

$$ma = F_{resultant} = F_{driving} - F_{friction} \checkmark$$

Apply to the two situations:

$$m \times 0.4 = (700 - F_{friction}) \quad \text{(Equation A)} \checkmark$$

$$m \times 0.8 = (1200 - F_{friction}) \quad \text{(Equation B)} \checkmark$$

Multiply Equation A by 2:

$$2(m \times 0.4) = 0.8m = 2 \times (700 - F_{friction})$$

Then substitute this into Equation B:

$$0.8m = 1200 - F_{friction} = 2 \times (700 - F_{friction})$$

Therefore:

$$1200 - F_{friction} = 1400 - 2F_{friction}$$

$$F_{friction} = 200\,N \checkmark$$

(ii) Substitute the value for friction into Equation A to determine the mass:

$$m \times 0.4 = (700 - F_{friction}) = 700 - 200 = 500$$

$$m = 1250\,kg \checkmark$$

ⓔ You can quickly check your answers by confirming that the substitution into Equation B gives the same answer.

Question 7

(a) (i) Define the linear momentum of an object. (1 mark)

(ii) Using continuous prose, state Newton's second law in terms of the momentum of a body. (2 marks)

(b) A ball is thrown at a wall so as to strike it at a right angle. The ball rebounds from the wall. The diagram below shows how the force F on the ball changes with time t during its contact with the wall.

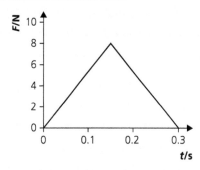

(i) Use the graph to calculate the change in momentum of the ball as a result of the collision. (4 marks)

(ii) The mass of the ball thrown at the wall is 60 g. If the collision is approximated to be elastic, find the magnitude of the initial velocity of the ball. (3 marks)

(c) A proton of mass 1.7×10^{-27} kg, moving at a speed of 4.0×10^6 m s^{-1}, collides with a stationary alpha particle of mass 6.8×10^{-27} kg and rebounds directly backwards. The alpha particle moves off along the same line with a velocity of 1.6×10^6 m s^{-1}. Show that the collision is elastic. (4 marks)

Answer to Question 7

(a) (i) Linear momentum is the product of mass and velocity. ✓

(ii) The rate of change of momentum of a body is equal to the (resultant) force applied ✓ and takes place in the direction of the force. ✓

(b) (i) Area under the triangle = $F \times \Delta t$ ✓

$F \times \Delta t$ = change in momentum ✓

area = $0.5 \times 8.0 \times 0.3$ ✓

momentum change = 1.2 N s or kg m s^{-1} ✓

(ii) In an elastic collision, KE is conserved, so the ball will come back with the same magnitude of velocity. ✓

initial momentum = 0.6 N s ✓

$p_{\text{initial}} = m \times v$

$v = \dfrac{p_{\text{initial}}}{m} = \dfrac{0.6}{0.06} = 10$ m s^{-1} ✓

ⓔ Note the need to convert the mass from g to kg to avoid a 10^n error.

(c) The diagram shows the collision.

Before 4.0×10^6 m s^{-1} Stationary

Proton   Alpha particle

v 1.6×10^6 m s^{-1}

After

ⓔ It is helpful to draw a simple diagram to depict the before and after with values included.

Use the conservation of momentum to find the velocity v of the rebounding proton:

momentum before collision = $1.7 \times 10^{-27} \times 4.0 \times 10^6 = 6.8 \times 10^{-21}$ kg m s^{-1}

momentum after collision = $(1.7 \times 10^{-27} \times v) + (6.8 \times 10^{-27} \times 1.6 \times 10^6)$ ✓

Momentum before = momentum afterwards, so:

$6.8 \times 10^{-21} = (1.7 \times 10^{-27} \times v) + (6.8 \times 10^{-27} \times 1.6 \times 10^6)$

$6.8 \times 10^{-21} = (1.7 \times 10^{-27} \times v) + (10.9 \times 10^{-21})$

$-4.1 \times 10^{-21} = 1.7 \times 10^{-27} \times v$

giving

$v = -2.4 \times 10^6$ m s^{-1} ✓

Then

initial KE $= \frac{1}{2} \times (1.7 \times 10^{-27}) \times (4.0 \times 10^6)^2 = 1.36 \times 10^{-14}$ J ✓

final KE $= \frac{1}{2} \times (1.7 \times 10^{-27}) \times (-2.4 \times 10^6)^2 + \frac{1}{2} \times (6.8 \times 10^{-27}) \times (1.6 \times 10^6)^2$

         $= 1.36 \times 10^{-14}$ J ✓

The KE is the same before and after the collision, therefore it is elastic.

Question 8

(a) Distinguish between kinetic energy and gravitational potential energy.   (2 marks)

(b) A particle possesses energy in two forms only: kinetic and gravitational potential energy. It has total energy 4.0 J and is initially at rest. Its potential energy E_p changes, causing a corresponding change in its kinetic energy E_k. No external work is done on or by the system.

Sketch a graph of E_p against E_k.

Explain how your graph illustrates the principle of the conservation of energy. (3 marks)

(c) A physics student enters the high jump event at the school sports. She estimates that she will need to raise her centre of mass by 1.6 m to clear the bar at a winning height. She will also have to move her centre of mass forwards horizontally at a speed of 0.80 m s^{-1} at the top of her jump in order to roll over the bar.

(i) The student's mass is 65 kg. Estimate the total energy required to raise her centre of mass and roll over the bar. (3 marks)

(ii) The student assumes that this energy can be supplied entirely from the kinetic energy she will have at the end of her run-up. Estimate the minimum speed she will require at the end of her run-up. (2 marks)

Answer to Question 8

(a) Kinetic energy E_k is energy possessed by a body because of its motion. ✓

Gravitational potential energy E_p is energy possessed by a body because of its position/height. ✓

(b) See graph.

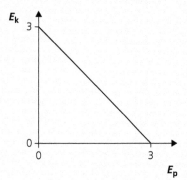

Straight-line graph through (0, 3) and (3, 0). ✓

Correctly labelled. ✓

The sum of $E_k + E_p$ is constant at all points on the line. ✓

(c) (i) $E_k = \frac{1}{2}mv^2 = 0.5 \times 65 \times 0.8^2 = 20.8\,\text{J}$ ✓

$E_p = mg\Delta h = 65 \times 9.81 \times 1.6 = 1020\,\text{J}$ ✓

total energy = 1041 J ✓

(ii) $\frac{1}{2}mv^2 = 0.5 \times 65 \times v^2 = 1041\,\text{J}$ ✓

$v = 5.7\,\text{m s}^{-1}$ ✓

Question 9

A 2 kW kettle is half filled with water and plugged into a 230 V mains supply. After 3 minutes the water is boiling and the kettle automatically switches off.

(a) **(i)** Calculate the heat energy dissipated by the element in the 3 minutes. (1 mark)

    (ii) Calculate the amount of charge that passes through the element of the kettle in this time. (2 marks)

    (iii) Hence calculate the number of electrons that pass through the element in this time. (2 marks)

(b) Calculate the resistance of the element. (2 marks)

Answer to Question 9

(a) **(i)** $W = P \times t = 2000 \times 180 = 360\,000\,\text{J}$ ✓

    (ii) $W = QV$ ✓

and

$$Q = \frac{W}{V} = \frac{360\,000}{230} = 1565\,\text{C} ✓$$

Alternatively:

$$P = VI \quad \text{or} \quad I = \frac{P}{V} = \frac{2000}{230} = 8.7\,\text{A} ✓$$

$$Q = It = 8.7 \times 180 = 1565\,\text{C} ✓$$

ⓔ Always remember to convert time into the base unit, seconds.

    (iii) $Q = N \times e$ ✓

$$N = \frac{Q}{e}$$

N = number of electrons

e = charge of one electron = $1.6 \times 10^{-19}\,\text{C}$

$$N = \frac{1565}{1.6 \times 10^{-19}} = 9.78 \times 10^{21} ✓$$

(b) $R = \frac{V}{I} = \frac{230}{8.7} = 26.4\,\Omega$ ✓✓

Alternatively:

$$P = \frac{V^2}{R} \quad \text{so} \quad R = \frac{V^2}{P} ✓$$

$$R = \frac{230^2}{2000} = 26.5\,\Omega ✓$$

ⓔ The value of R obtained is slightly different owing to using rounded values of I. Either answer is acceptable.

Question 10

(a) A piece of wire is stretched so that its length is increased by 1%, but its volume remains unchanged.

    (i) How will the resistivity of the wire be affected, if at all?   (1 mark)

    (ii) How will the resistance of the wire be affected, if at all?   (2 marks)
        Quantify any changes that occur.

(b) The circuit below shows two different 6.0 V bulbs powered by the same battery. Bulb A is rated as 2.4 W and bulb B as 4.8 W.

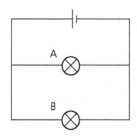

    (i) Calculate the current delivered by the battery.   (2 marks)

    (ii) Calculate the resistance of the filament of bulb A.   (1 marks)
        The filament of bulb A is constructed from wire of cross-sectional area $2.0 \times 10^{-8}\,m^2$ and is made of material of resistivity $7.5 \times 10^{-6}\,\Omega\,m$.

    (iii) Use your answer in part (ii) to calculate the length of the filament of bulb A.   (2 marks)

    (iv) Suggest how this length of wire may be fitted into a small light bulb.   (1 mark)

Answer to Question 10

(a) (i) Resistivity is a property of the material, which has not changed, so the resistivity is unchanged. ✓

    (ii) Resistance is dependent on the dimensions of the wire, which have changed (longer and thinner), so the resistance will change. ✓

$$R = \frac{\rho L}{A}$$

Length L increased by 1%. Cross-sectional area A will decrease by 1%, as volume is constant and $V = L \times A$.

Consequently, R will increase by 2%. ✓

(b) (i) $P_{total} = P_A + P_B = 2.4 + 4.8 = 7.2\,W$ ✓

$$P = VI \qquad \text{or} \qquad I = \frac{P}{V} = \frac{7.2}{6} = 1.2\,A \checkmark$$

Alternatively, you can work out the current for each bulb and add them:

$$I_A = \frac{P}{V} = \frac{2.4}{6.0} = 0.4\,A$$

$$I_B = \frac{P}{V} = \frac{4.8}{6.0} = 0.8\,A \checkmark$$

$$I_{total} = I_A + I_B = 0.4 + 0.8 = 1.2\,A \checkmark$$

(ii) $R = \dfrac{V}{I}$

$R_A = \dfrac{6.0}{0.4} = 15\,\Omega$ ✓

(iii) $R = \dfrac{\rho L}{A}$ ✓

Rearranging:

$L = \dfrac{RA}{\rho} = \dfrac{15 \times 2.0 \times 10^{-8}}{7.5 \times 10^{-6}} = 4 \times 10^{-2}\,\text{m or } 4\,\text{cm}$ ✓

(iv) Coiled, but not touching. ✓

ⓔ Check the specification. If it refers to 'recall and use', a mark may be awarded for simply recording the equation, as it will not appear on the data and formulae sheet. In the case of $R = V/I$ in part (b)(ii), no mark was given, as this has already been awarded in question 9(b) in this test.

Question 11

Describe an experiment to measure the internal resistance of a 1.5 V battery. Include the following in your answer:

- **a circuit diagram**
- **the procedure followed**
- **the measurements taken**
- **how the measurements are used**
- **how safety is managed**

(7 marks)

ⓔ In most papers, there will be a question based on an experiment specified in your specification. Generally, space will be left for a labelled diagram. Use a ruler and pencil. Follow the instructions for the question but give a clear step-by-step account of the method, stating the apparatus used and all readings taken. Always address accuracy and reliability.

Answer to Question 11

Correct diagram. ✓✓

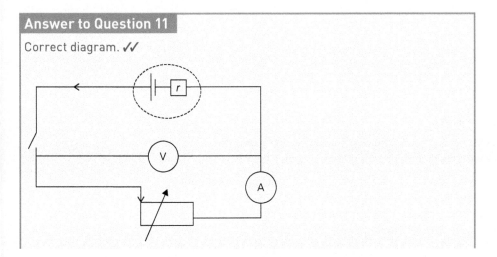

1   Connect the circuit as shown, with the variable resistor at its maximum value setting. This is the external load resistance.

2   Close the switch and immediately record the readings of current and terminal p.d. from the ammeter and voltmeter. ✓

3   Open the switch as soon as the values are recorded to reduce discharging the battery needlessly.

4   Move the slider on the variable resistor to decrease the load resistance, and repeat steps 2 and 3.

5   Repeat the procedure of reducing the load resistance several times until its value is near zero. ✓

     It is important to minimise the time for which the switch is closed when the load is near zero, because the current will be high and the battery could be damaged. ✓

6   Plot a graph of p.d. on the y-axis against current on the x-axis. ✓

7   Measure the *magnitude* of the slope of the graph, which is equal to the internal resistance of the battery:

     $V_{terminal} = -Ir + E$ and $y = mx + c$

     gradient, $m = -r$, internal resistance ✓

Question 12

(a) Copy the axes shown below and sketch graphs to show the variation with temperature of the resistance of:

   (i)   a pure metallic conductor   (2 marks)
   (ii) a negative-temperature-coefficient (NTC) thermistor   (2 marks)

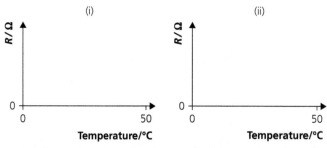

   (iii) In terms of molecular theory, explain why the resistances of both components vary with temperature in the manner you have sketched.   (4 marks)

(b) Superconductivity is a phenomenon observed in several metals and ceramic materials. With the help of a sketch graph, explain what is meant by superconductivity. Label the axes with both quantities and appropriate units. Also label the significant feature of the plot.   (5 marks)

Answer to Question 12

(a) See graphs.

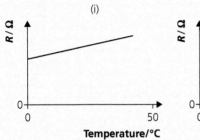
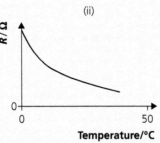

(i)

(ii)

(i) non-zero start ✓; increasing linear or slight curve ✓

(ii) non-zero start ✓; decreasing curve but not cutting axis ✓

(iii) For the metal: as the temperature increases, the positive ion vibration is greater and the interaction between electrons and ions is more substantial — hence greater resistance. ✓✓

For the thermistor: this mechanism also occurs but is more than offset by the increase in the number of conduction electrons as the temperature increases — so resistance decreases. ✓✓

(b) Superconductivity is the condition displayed in some materials of an *abrupt* drop to zero resistance as they are cooled to temperatures ranging from near absolute zero (0 K) to about 100 K. ✓✓

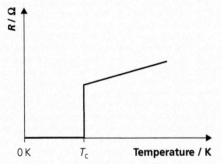

0 K T_c **Temperature / K**

T_c = transition or critical temperature

Axes: R/Ω and T/K ✓

Shape of sketch ✓

Transition temperature labelled ✓

Total: 100 marks

■Self-assessment test 2

Question 1

(a) Six physical quantities are listed below. Select which of the quantities are **vectors** and state their SI base units. (4 marks)

kinetic energy   frequency   velocity   charge   force   power

(b) The radius of the hydrogen nucleus is 1 fm. Express this value in nm, μm and m. (3 marks)

Answer to Question 1

(a) velocity, ✓ $m\,s^{-1}$; ✓

force, ✓ $kg\,m\,s^{-2}$ ✓

(b) $1 \times 10^{-6}\,nm$ ✓

$1 \times 10^{-9}\,\mu m$ ✓

$1 \times 10^{-15}\,m$ ✓

Question 2

The graph below shows the variation in the displacement s with time t for a car of mass 1200 kg, which is travelling in a straight line. After $t = 2\,s$, the car's brakes are applied and the car decelerates uniformly to rest, stopping at time $t = 5\,s$.

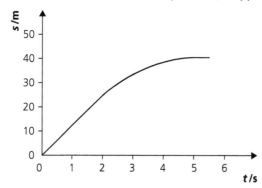

Determine each of the following quantities, showing clearly in each case how you obtain your answer:

(a) the speed of the car at time $t = 2\,s$ (1 mark)

(b) the value of the deceleration (2 marks)

(c) the force required to produce the deceleration (2 marks)

(d) the distance travelled by the car while slowing down (1 mark)

(e) the work done in stopping the car (2 marks)

Answer to Question 2

(a) The graph is a straight line up from $t = 2$ s.

The slope of the displacement–time graph equals the velocity, which is the same as the speed because it is in a straight line up to this point.

$$\text{speed} = \frac{\Delta s}{\Delta t} = \frac{24}{2} = 12\,\text{m s}^{-1} \checkmark$$

(b) Acceleration is given by:

$$a = \frac{\Delta v}{\Delta t} = \frac{v - u}{t} \checkmark$$

$$a = \frac{0 - 12}{5 - 2} = -4\,\text{m s}^{-2}$$

deceleration $= 4\,\text{m s}^{-2} \checkmark$

ℯ The negative should be dropped from the answer line because the question asks for deceleration not acceleration — the word incorporates the negative.

(c) $F_{\text{resultant}} = ma$ ✓

$F_{\text{resultant}} = 1200 \times (-4) = -4.8\,\text{kN}$ (the negative indicates direction) ✓

(d) From the graph we can read off the displacement values at $t = 2$ s and $t = 5$ s:

$\Delta s = 42 - 24 = 18\,\text{m}$ ✓

(e) work done $=$ force $\times$ distance moved in the direction of the force ✓

work done $= -4800 \times 18 = (-)86\,400\,\text{J}$ ✓

ℯ The negative value for work done is a consequence of the directions of the displacement and the force being opposite. As work is a scalar, direction is irrelevant and we deal with only its magnitude. In this case, work has been done *on* the force, rather than *by* the force.

Question 3

(a) A car of mass 950 kg is moving up a road inclined at 12° to the horizontal with an acceleration of 0.50 m s^{-2}. At a particular instant, its speed is such that the opposing force of friction experienced by the car is 510 N. Calculate the driving force when the car is moving at this speed. (5 marks)

(b) The maximum speed of the car up this slope is 18 m s^{-1}. When moving at this speed, the opposing force of friction has increased to 650 N.

   (i) Calculate the driving force when moving at the constant speed of 18 m s^{-1}. (3 marks)

(ii) What is the mechanical power produced by the car at this speed? (2 marks)

(iii) If, at this speed, the car engine uses fuel energy at a rate of 180 kW, determine the percentage efficiency of the engine at this speed. (2 marks)

(c) The car is now brought to rest on the inclined road. The engine is switched off and the hand brake is applied, but unfortunately it does not prevent the car from moving. After a short interval, the car moves slowly with constant velocity down the slope. Explain how the principle of conservation of energy applies as the car moves down the slope with constant velocity. (3 marks)

Answer to Question 3

(a) Force down the incline = $F_{friction} + mg\sin\theta$ ✓

$F_{resultant} = F_{driving} - (F_{friction} + mg\sin\theta)$ ✓

$F_{resultant} = ma$ ✓

$F_{driving} = (950 \times 0.50) + (510 + [950 \times 9.81\sin 12°])$ ✓

$F_{driving} = 2900\,N$ (to 2 s.f.) N ✓

(b) (i) At maximum speed, there is no acceleration. ✓

$F_{resultant} = 0$ so $F_{driving} = F_{friction} + mg\sin\theta$ ✓

$F_{driving} = 650 + (950 \times 9.81\sin 12°)$

$F_{driving} = 2600\,N$ (to 2 s.f.) N ✓

ⓔ The answer is given to two significant figures as the data provided are to two significant figures.

(ii) $P = Fv$ ✓

$P = 2600 \times 18 = 47\,kW$ ✓

(iii) Efficiency is given by:

$\text{efficiency} = \dfrac{\text{power output}}{\text{power input}}$ ✓

$\text{efficiency} = \dfrac{47}{180} = 0.26$

percentage efficiency = $0.26 \times 100 = 26\%$ ✓

(c) The car's potential energy decreases as it moves down the slope. ✓

The car moves with constant velocity, which means that its kinetic energy is not changing. ✓

So the change in potential energy must equal the energy dissipated as heat in the brakes ✓ (so conservation applies).

Question 4

(a) Define the moment of a force about a point. (2 marks)

(b) State the principle of moments. (3 marks)

(c) When playing long shots, snooker players often use a cue support known as a rest.

The snooker cue itself is a non-uniform rod of length 150 cm and mass 0.57 kg.

If the cue is placed on the rest 40 cm from the tip, as shown in the diagram below (which is not to scale), the player has to apply a force of 3.0 N at the other end to balance the cue.

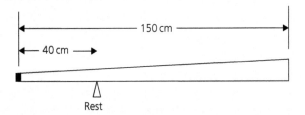

(i) Determine the distance of the centre of mass of the cue from the tip. (4 marks)

(ii) What is the support force provided by the rest? (2 marks)

Answer to Question 4

(a) The moment of a force about a point is the product of the force and the *perpendicular* ✓ distance of the line of action of the force from the point. ✓

(b) When an object is in rotational equilibrium, ✓ the sum of the clockwise moments about a point is equal to the sum of the anticlockwise moments ✓ about the same point. ✓

e Be careful to include all the points in definitions, to access all the marks allocated. This can vary from paper to paper — for example, the definition in part (a) has been allocated a mark for 'perpendicular'.

(c) Correct diagram ✓

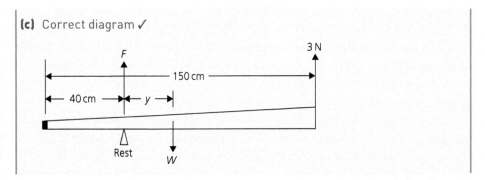

e With moments questions such as this, it is essential to draw a good labelled diagram.

(i) Take moments about the 'rest' (pivot). Let the distance to the centre of gravity be y. Then:

clockwise moments = $W \times y = (0.57 \times 9.81) \times y$

anticlockwise moments = $3 \times (150 - 40)$

For rotational equilibrium:

$(0.57 \times 9.81) \times y = 3 \times (150 - 40)$ ✓

$$y = 3 \times \frac{150 - 40}{0.57 \times 9.81}$$

$y = 59\,\text{cm}$ ✓

ⓔ Make sure you finish the calculation — this is not what the question asked for.

So the distance from the tip to the centre of gravity is 99 cm. ✓

ⓔ Note the use of centimetres rather than converting to metres. This is an acceptable sub-multiple, but must be used consistently throughout the question. Although the answer is technically given to three significant figures, because the addition takes us beyond 100, it makes sense here, as the data given were to the nearest centimetre.

(ii) For vertical equilibrium:

$F_{\text{resultant}} = 0$ or $F_{\text{up}} = F_{\text{down}}$ ✓

Therefore:

$3 + F = W = (0.57 \times 9.81)$

$F = 2.6\,\text{N}$ ✓

Question 5

(a) A cricketer strikes a cricket ball of mass 0.15 kg at right angles to the bat so that the direction of the ball is reversed. The ball is travelling horizontally at $22\,\text{m s}^{-1}$ just before it is struck by the bat, and leaves with a speed of $38\,\text{m s}^{-1}$, as shown in the diagram below.

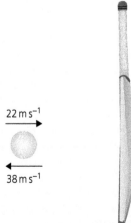

(i) What is the magnitude of the impulse exerted on the ball by the bat? (4 marks)

(ii) If the bat and ball are in contact for 0.01 s, what is the average force exerted on the ball by the bat? (3 marks)

(iii) The cricketer extends her follow through so that the bat and ball are in contact for 0.015 s. Assuming the force remains the same, at what speed will the ball leave the bat? (2 marks)

(b) A bullet of mass 16 g is fired horizontally from a gun with a velocity of 280 m s⁻¹. It hits, and becomes embedded in, a block of wood of mass 3.0 kg, which is freely suspended by strings, as shown in the diagram below. Air resistance is negligible.

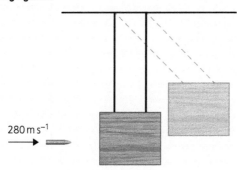

280 m s⁻¹

(i) Calculate the magnitude of the momentum of the bullet as it leaves the gun. Give your answer in N s. (2 marks)

(ii) Calculate the magnitude of the initial velocity of the wooden block and bullet after impact. (2 marks)

(iii) Use your answer to (ii) to calculate the kinetic energy of the wooden block and embedded bullet immediately after impact. (2 marks)

(iv) Show by calculation whether the collision of bullet and block is elastic or inelastic in type. (3 marks)

(v) Calculate the maximum height above the equilibrium position to which the wooden block, with the embedded bullet, rises after impact. (3 marks)

Answer to Question 5

(a) (i) impulse = $mv - mu$ ✓

         Using + for u and – for v to indicate opposite directions: ✓

    impulse = 0.15([–38] – [+22]) ✓

    = 0.15 × –60 = (–)9.0 kg m s⁻¹ ✓

ⓔ You are asked for the magnitude, so the negative sign is dropped.

    (ii) impulse = Ft ✓

       –9.0 = F × 0.01 ✓

       F = –900 N ✓

ⓔ The negative sign indicates the direction of the force on the ball by the bat.

Questions & Answers

(iii) $Ft = mv - mu$

$mv = Ft + mu$

$v = \dfrac{Ft + mu}{m} = \dfrac{[-900 \times 0.015] + [0.15 \times 22]}{0.15}$ ✓

$v = -68\,\text{m s}^{-1}$ ✓

ⓔ The second mark is awarded if the signs are correct.

(b) (i) $p = m \times u$

$p = 0.016 \times 280$ ✓

$p = 4.5\,\text{kg m s}^{-1}$ or N s ✓

(ii) momentum after $= (m_1 + m_2)v$

momentum before = momentum after

$4.5 = (0.016 + 3)v$ ✓

$v = 1.5\,\text{m s}^{-1}$ ✓

(iii) $KE = \frac{1}{2}mv^2$ ✓

$KE = \frac{1}{2} \times 3.016 \times 1.5^2 = 3.3\,\text{J}$ ✓

(iv) $KE_{\text{bullet}} = \frac{1}{2} \times 0.016 \times 280^2 = 630\,\text{J}$ ✓

Therefore KE is not conserved. ✓
Hence the collision is inelastic. ✓

(v) Conservation of energy applied, transfer from KE to GPE:

$\Delta KE = \Delta GPE$ ✓

$\frac{1}{2}mv^2 - \frac{1}{2}mu^2 = mg\Delta h$

$3.3 = 3.016 \times 9.8 \times \Delta h$ ✓

$\Delta h = 0.11\,\text{m}$ ✓

Question 6

A person playing darts stands on a horizontal floor, 3.0 m from the dart board, which hangs vertically. A dart thrown towards the board leaves the player's hand at a height of 1.8 m above floor level, and initially travels horizontally. The dart strikes the board 1.5 m above floor level. Air resistance is negligible.

(a) Sketch graphs to show how:

(i) the kinetic energy of the dart

(ii) the gravitational potential energy of the dart, relative to the floor

vary with time, from the instant the dart is thrown until it strikes the board. **(4 marks)**

(b) (i) Calculate the time of flight of the dart. (3 marks)

(ii) Calculate the initial speed of the dart. (2 marks)

(iii) Find the magnitude and direction of the velocity of the dart as it enters the board. (7 marks)

Answer to Question 6

(a) (i)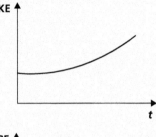

positive intercept ✓
upward curve ✓

(ii)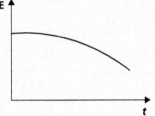

positive intercept ✓
downward curve ✓

(b) (i) $s = ut + \frac{1}{2}at^2$ ✓

$s = 1.8 - 1.5 = 0.3$ ✓

$0.3 = 0 + 0.5 \times 9.81 \times t^2$

$t = 0.25\,\text{s}$ ✓

(ii) Horizontal component is constant. ✓

$\text{speed} = \dfrac{\text{distance}}{\text{time}} = \dfrac{3.0}{0.25} = 12.0\,\text{m s}^{-1}$ ✓

(iii) For the vertical component of velocity at the target:

$v_v = u_v + gt$ or $s = \frac{1}{2}(u_v + v_v)t$ ✓

$v_v = 0 + 9.8 \times 0.25 = 2.45 = 2.5\,\text{m s}^{-1}$ ✓

Appreciates the vector addition of v_h and v_v ✓

magnitude of velocity $= \sqrt{12.0^2 + 2.45^2}$

$= \sqrt{150} = 12.25 = 12.3\,\text{m s}^{-1}$ ✓

direction: $\tan\theta^{-1} = \dfrac{2.45}{12.25} = 0.2$ ✓

$\theta = 11.3°$ ✓

Below the horizontal, or shown in a sketch ✓

Question 7

(a) Electrical conductors may be classified as **ohmic** or **non-ohmic**.

    (i) State Ohm's law.   (2 marks)

    (ii) Name one example of an **ohmic** conductor.   (1 mark)

(b) The current–voltage graph for a certain device is shown below.

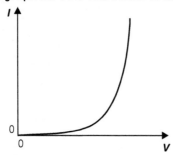

    (i) State the name of the device.   (1 mark)

    (ii) State whether this device is an ohmic conductor or a non-ohmic conductor. Explain your answer, making reference to the graph.   (2 marks)

Answer to Question 7

(a) (i) For a conductor under constant physical conditions *or* at constant temperature, ✓ current is proportional to potential difference. ✓

    (ii) *Named* metal, e.g. copper, tungsten ✓

ⓔ Metal may be accepted but is not the best answer.

(b) (i) (Semiconductor) diode ✓

ⓔ Examiners do not like 50/50 questions like this, but they do still appear as an unavoidable lead-in.

    (ii) The device is a non-ohmic conductor ✓ because the graph is not a straight line through the origin. ✓

ⓔ 'Because current is not proportional to the potential difference' is factually correct. However, examiners *may* take the view that this does not make direct reference to the graph.

Question 8

(a) Three 6 Ω resistors are arranged as shown in the circuit diagram below. What is the effective resistance between A and B?   (4 marks)

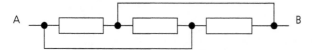

(b) The circuit shown below is constructed of resistors, each of which has a maximum power rating of 1 W. Calculate the maximum potential difference that can be safely applied across AB without damage to any resistor. You must show your working clearly.

(6 marks)

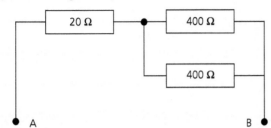

Answer to Question 8

(a)

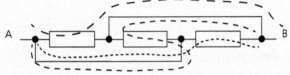

Identify two different routes for the current. ✓

Identify a third different route for the current. ✓

Effectively three 6 Ω resistors in parallel. ✓

$$\frac{1}{R_{\text{effective}}} = \frac{1}{R_1} + \frac{1}{R_2} + \frac{1}{R_3} = \frac{3}{6}$$

$R_{\text{effective}} = 2\,\Omega$ ✓

(b) Need to identify which resistor would reach the maximum power first.

Recognise the current I splitting equally through the 400 Ω resistors. ✓

Using $P = VI = I^2R = \dfrac{V^2}{R}$: ✓

for 20 Ω resistor, $P = I^2 \times 20 = 20I^2$

for 400 Ω resistor, $P = \left(\dfrac{1}{2}\right)^2 \times 400 = 100I^2$

Therefore the power usage is five times greater in the 400 Ω resistors. So we must set the safe p.d. based on their limits. ✓

$P = 1\,\text{W}; R = 400\,\Omega$

$P = \dfrac{V^2}{R} \qquad V^2 = PR = 1 \times 400$

Therefore, for the 400 Ω resistor: $V = 20\,\text{V}$ ✓

Effective resistance of the two 400 Ω resistors in parallel is 200 Ω. ✓

Therefore the p.d. drop across the 20 Ω is 20/10 = 2 V.

The maximum p.d. that can be applied safely is 22 V. ✓

🄔 Both parts of this difficult question would be classed as unstructured, so you must show each stage of your attempted solution and the relevant equations. The equations and components of the solution will draw independent marks.

Question 9

(a) The diagram below shows three wires meeting at a junction in a circuit.

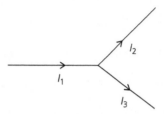

(i) The wires carry currents I_1, I_2 and I_3, as shown. Write down the equation relating all three currents.   (1 mark)

(ii) If the values of I_1 and I_2 are both 3 A, what is the value of I_3?   (1 mark)

(b) The sensor circuit shown below provides a voltage V that depends on the brightness of the lighting in a room.

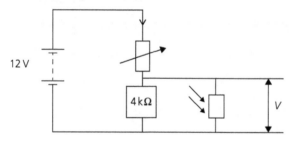

The battery has an e.m.f. of 12 V and negligible internal resistance.

The light sensor is a light-dependent resistor (LDR). The resistance of the LDR is 5.0 kΩ in the dark and 0.20 kΩ when the natural light in the room is at its brightest.

A voltage $V = 3$ V is required to trigger the lighting circuit.

The variable resistor R acts as a sensitivity control for the circuit.

Find the value of R that will cause the lighting circuit to activate when the LDR has a resistance of 4 kΩ, just when it is getting dark.   (4 marks)

Answer to Question 9

(a) (i) $I_1 = I_2 + I_3$ ✓

(ii) $3 = 3 + I_3$

$I_3 = 0\,\text{A}$ ✓

(b) To activate when the LDR has resistance of $4\,\text{k}\Omega$:

$$\frac{1}{R_{\text{effective}}} = \frac{1}{R_1} + \frac{1}{R_2}\ ✓$$

$$\frac{1}{R_{\text{effective}}} = \frac{1}{4} + \frac{1}{4}$$

$R_{\text{effective}} = 2\,\text{k}\Omega$ ✓

Amended version of potential divider equation:

$$V_{\text{out}} = \frac{R_{\text{effective}} V_{\text{in}}}{R_{\text{effective}} + R}\ ✓$$

$$3 = \frac{2 \times 12}{2 + R}$$

$$(2 + R) = \frac{2 \times 12}{3}$$

$R = 6\,\text{k}\Omega$ ✓

Total: 100 marks

Knowledge check answers

1 A unit from which other quantities can be derived.
2 s, m, kg, A, K, mol, (cd)
3 2.4×10^{-8} m
4 work, scalar; density, scalar; charge, scalar; momentum, vector
5 Scalar: speed, time, distance, gravitational potential energy, kinetic energy
   Vector: velocity, displacement, acceleration, weight, momentum
6 a 7N   b 17N   c 13N
7 Pushing is more effective because the vertical component will act downwards to add to the flattening effect.
8 horizontal: $10.4\,\mathrm{m\,s^{-1}}$, vertical: $6.0\,\mathrm{m\,s^{-1}}$
9 a 40N   b 20N
10 87.2N
11 50m
12 $5.9\,\mathrm{m\,s^{-2}}$
13 12.5m
14 120m, $2.5\,\mathrm{m\,s^{-2}}$
15 5.7s
16 Parabola

17 a $a = -g$, $u_v = 5.0\,\mathrm{m\,s^{-1}}$, $u_h = 8.7\,\mathrm{m\,s^{-1}}$
   b $a = -g$, $u_v = 0\,\mathrm{m\,s^{-1}}$, $u_h = 8.7\,\mathrm{m\,s^{-1}}$
18 3.4s (symmetry)
19 9810N
20 600N
21 440N
22 $500\,\mathrm{kg\,m\,s^{-1}}$
23 $[\mathrm{N}] = \mathrm{kg\,m\,s^{-2}}$
   $[\mathrm{N\,s}] = \mathrm{kg\,m\,s^{-1}}$
24 $0.5\,\mathrm{m\,s^{-1}}$
25 2.3×10^5 J
26 $6.3\,\mathrm{m\,s^{-1}}$
27 24000 J
28 75%
29 a 90C   b 5.6×10^{20}
30 a 12.5A   b 2.7MJ
31 $4.0\,\Omega$
32 96J
33 a $36\,\Omega$ b $4\,\Omega$
34 $\rho = 1.1 \times 10^{-6}\,\Omega\,\mathrm{m}$
35 a When delivering current.
   b When no current is being drawn.
   c When being charged by an external source.

Note: **bold** page numbers indicate defined terms.

A

acceleration 17
   of freefall 21–24, 29

B

base units 6

C

car safety 33
centre of mass/gravity 14–15
circuit diagrams 47–49
command terms 66
conservation of charge 46
conservation of energy **40**–42, 46
conservation of momentum **35**–39
critical temperature 58
current 44–45
current–voltage characteristics 53–55

D

data and formulae sheets 67
derived units 6–7
displacement–time graph 19
dynamics, projectile motion 25–26

E

efficiency **42**–43
elastic collisions 39
electrical conductors 55–58
electrical resistance 47
electric current **44**–45
electromotive force **46**
equations
   homogeneous 7
   of motion 18–19
   and units 7
external forces, linear momentum 35

F

force–time graphs 34
frictional forces 27

G

Galileo 29

gravitational potential energy **40**
gravity 29

H

heater circuit 64
homogeneous equations 7

I

impulse **32**, 34
inelastic collisions, linear momentum 39
insulators 55
internal resistance 61
International System of Units (SI units) 6–7

K

kinetic energy 40
Kirchhoff's laws 46

L

lighting circuit 65
linear momentum
   car safety 33
   conservation of momentum **35**–39
   elastic collisions 39
   and external forces 35
   'follow through' 33–34
   impulse 32
   inelastic collisions 39
   momentum 31
   Newton's second law 31–32
linear motion
   acceleration of freefall 21–24
   displacement–time graph 19
   equations 18–19
   quantities 16–18
   velocity–time graphs 20

M

moment of a force **12**–16

N

Newton's laws of motion 27–31, 31–32, 35–36
NTC thermistors 56–58

O

Ohm's law **55**–58

P

parallel circuits 48–50
physical quantities 6–9
potential difference **45**
potential divider circuits 61–65
power **42**
prefixes, base units 7
principle of moments **12**–16
projectile motion **25**–26
Pythagoras' theorem, and vectors 9

R

resistance **47**
resistivity **50**–53
revision tips 67

S

scalars 9
scale drawing, and vectors 10
self-assessment tests
   test 1 68–81
   test 2 82–93
series circuits 47–48
SI base units 6
SI derived units 6–7
SI units (International System of Units) 6–7
speed 17
superconductivity 58

T

thermistors 55–58

V

variable potential 62
vectors 9–12
velocity 17
velocity–time graphs 20
vertical equilibrium 15–16

W

weight 29
work done 39–40